Project Planning and Project Success

The 25% Solution

Best Practices and Advances
in Program Management Series

Series Editor
Ginger Levin

RECENTLY PUBLISHED TITLES

Project Planning and Project Success: The 25% Solution
Pedro Serrador

Project Health Assessment
Paul S. Royer, PMP

Portfolio Management: A Strategic Approach
Ginger Levin and John Wyzalek

Program Governance
Muhammad Ehsan Khan

Project Management for Research and Development:
Guiding Innovation for Positive R&D Outcomes
Lory Mitchell Wingate

The Influential Project Manager:
Winning Over Team Members and Stakeholders
Alfonso Bucero

PfMP® Exam Practice Tests and Study Guide
Ginger Levin

Program Management Leadership: Creating Successful Team Dynamics
Mark C. Bojeun

Successful Program Management: Complexity Theory, Communication,
and Leadership
Wanda Curlee and Robert Lee Gordon

From Projects to Programs: A Project Manager's Journey
Samir Penkar

Sustainable Program Management
Gregory T. Haugan

Leading Virtual Project Teams: Adapting Leadership Theories
and Communications Techniques to 21st Century Organizations
Margaret R. Lee

Applying Guiding Principles of Effective Program Delivery
Kerry R. Wills

Project Planning and Project Success

The 25% Solution

Pedro Serrador, PhD

CRC Press
Taylor & Francis Group
Boca Raton   London   New York

CRC Press is an imprint of the
Taylor & Francis Group, an **Informa** business
AN AUERBACH BOOK

CRC Press
Taylor & Francis Group
6000 Broken Sound Parkway NW, Suite 300
Boca Raton, FL 33487-2742

© 2015 by Taylor & Francis Group, LLC
CRC Press is an imprint of Taylor & Francis Group, an Informa business

No claim to original U.S. Government works

Printed on acid-free paper
Version Date: 20150209

International Standard Book Number-13: 978-1-4822-0552-7 (Hardback)

Library of Congress Cataloging-in-Publication Data

Serrador, Pedro.
    Project planning and project success : the 25% solution / Pedro Serrador.
        pages cm. -- (Best practices and advances in program management series)
    Includes bibliographical references and index.
    ISBN 978-1-4822-0552-7 (hardcover : alk. paper) 1. Project management. 2. Strategic planning. I. Title.

    HD69.P75S43 2015
    658.4'04--dc23
                                                        2014039345

Visit the Taylor & Francis Web site at
http://www.taylorandfrancis.com

and the CRC Press Web site at
http://www.crcpress.com

Contents

Introduction

If you don't know where you are going, you'll end up some place else.

Yogi Berra

As someone who has been managing and consulting on projects for 25 years, the aspects of what make a successful project have been of obvious interest. After all, delivering successful projects is how I made my living. Other project managers no doubt feel the same.

When I took my life in a more academic direction, I was, of course, interested in the question of what maximized success. I thought, "In the projects I had managed, how could I have avoided the problems?" I wondered whether planning and analysis were the basis of project success. This became the topic of my PhD thesis. This book grew out of the studies and research I did for my thesis.

For many years, I organized canoe trips to Canadian provincial parks such as Algonquin Park. I started this before I became a project manager and I think I learned a great deal from the experience. I would take a group of 8–18 people, most of them strangers, and lead them on a 3- or 4-day trip. Canoe trips into the backcountry, as it is called, are not easy to organize. You spend 5 hours or more canoeing into a lake. If you forget the matches or the salt for dinner, you can't drive to the corner store; you have to canoe 5 hours to get to the store and then canoe back. Unless you are lucky enough to meet other campers who are willing to share, you're out of luck.

For example, I found that for a typical group, 2.5–3 servings of bread per person per day was enough, on average, to cover lunches and breakfast. This was important as too much and you would have to find a place to store all this food. Worst of all, you end up "portaging" (i.e., carrying overland past waterfalls, etc.) extra food that would end up being thrown away. Too little and then you had hungry people to deal with. I remember on one trip while we were hiking up a hill we ran into a dad taking his kids canoeing. He was asking everyone he met if they had extra food. He was obviously not an experienced canoeist and did not pack enough food. We only had snacks with us on the hike but we gave him what we had.

Lots of things can go wrong on canoe trips. It rains and you don't have a raincoat or rain gear. You forget those matches and can't start a fire to cook your dinner or get warm. You plan a trip that is too long or with too much portaging and you can't make it to camp before nightfall. You forget the map and get lost. All of these will lead to problem trips and unhappy campers!

Here are some things I learned:

- Plan ahead. Something as challenging as a backcountry trip can't be successful if you just wing it. Running out of food and rationing people to a quarter sandwich and one-eighth of an apple for lunch is not fun!
- Find a method that works and stick to it. Tweaks are OK; trying something completely new is risky.
- People are the biggest variable. You never know how they will react in a given situation. You can never anticipate reactions, only deal with them in a reasonable way.
- Be flexible. Unexpected challenges and opportunities will arise. If you paddle past that perfect beach, better stop and go for a swim. You might not see another one.

Two of the things I learned were about careful planning, two were about flexibility.

And this is what I found in my research. Planning ahead is clearly and demonstrably important for project success. And being flexible (or agile) during execution will also lift the success of your project.

The rest of the book gets into the details of why those areas are important to managing critical projects. As well as to leading canoe trips.

1

Spectacular Project Failures

Everybody has a plan until they get punched in the face.

Mike Tyson

It is often said that some of the most notorious project failures could have been prevented by better work up front. You likely have examples from your own career. There are some well-known examples of project failures that could arguably have been avoided by better planning and analysis.

MARS CLIMATE ORBITER

The Mars Climate Orbiter (MCO) was launched in late 1998, followed by the Mars Polar Lander (MPL) and Deep Space 2 launched in early 1999. MCO failed to reach the orbit of Mars because of a navigational error. This resulted in the ship entering the atmosphere of Mars, and burning up instead of going into the planned orbit.

The following is a summary of the results of the MCO investigation. Spacecraft operational data needed for navigation were provided to the Jet Propulsion Laboratory (JPL) navigation team by the prime contractor Lockheed Martin in English units instead of metric units as specified in the contracts. This was the direct cause of the failure as it caused the craft to miscalculate the trajectory and burn up in the atmosphere. However, it is important to recognize that space missions are a "one strike and you're out" activity. Thousands of functions can be performed correctly and a single error can be catastrophic to the whole mission. Errors are prevented by planning, analysis, oversight, and independent testing, which were not good enough on the MCO project.

Specifically, software testing was inadequate. Equally important, the navigation team was not vigilant enough, did not understand the spacecraft, and was poorly trained. Anomalies in navigation (caused by the same error in units) observed during the trip from Earth to Mars were not adequately pursued to determine the cause; the opportunity to make a trajectory correction maneuver when nearing Mars was not used due to inadequate preparation. The MCO project may have had competent managers, but may have had an inexperienced project team. The lack of involvement of senior management to help compensate for this lack of experience may have contributed to the MCO failures.

The selection of a launch vehicle with little margin for error, growth in the scientific payload, and a fixed planetary launch window also contributed to inadequate margins. The result was deficiencies in the analysis, as well as inadequate testing in the plan before mission operations. These problems resulted in excessive risk and contributed to the failures. The net result was a multimillion-dollar spaceship uselessly burning up in the atmosphere of Mars.

FOXMEYER ENTERPRISE RESOURCE PLANNING (ERP) PROGRAM

In 1993, FoxMeyer Drugs was the fourth largest distributor of pharmaceutical products in the United States, worth approximately $5 billion. FoxMeyer bought an SAP system and a system of warehouse automation in an attempt to increase efficiency. It hired Andersen Consulting to integrate and implement both systems. It was supposed to be a project with a budget of $35 million.

There were several reasons for failure. First, FoxMeyer created an overly aggressive timeline: the whole system would be implemented in 18 months. Second, warehouse employees whose jobs were threatened by the automated system were not consulted prior to the project implementation. After three existing warehouses were closed, the first warehouse to be automated was plagued by sabotage, with inventory damaged by workers and orders not being filled. Finally, the new system was less effective than the one that it replaced. In 1994, SAP was processing only 10,000 requests per night, compared to 420,000 orders under the old mainframe.

In 1996, the company went bankrupt and was eventually sold to a competitor for only $80 million. In 1998, FoxMeyer sued Andersen and SAP for $500 million each, claiming that they had paid twice the estimate for a system with a fraction of the capabilities.

CANADA'S LONG-GUN REGISTRATION SYSTEM

In June 1997, Electronic Data Systems and SHL Systemhouse, at that time headquartered in the United Kingdom, began working on a system for a Canadian national gun registry. The original plan was for a modest information technology (IT) project that would cost only $2 million: $119 million for implementation, minus $117 million in user fees (all costs Canadian dollars).

But then politics got in the way. The pressure of the gun lobby and other interest groups resulted in more than 1,000 change requests in just the first two years. The changes involved interfacing the computer system with more than 50 agencies, and because the integration was not part of the original contract, the government had to pay for all the extra work. In 2001, the cost rose to $688 million, including $300 million for support.

But things got worse. In 2001, the annual maintenance costs were running at $75 million per year. By 2002 an audit estimated that the program would end up costing more than $1 billion by 2004, while generating revenue of just $140 million, giving rise to the nickname given to it by its critics: "Billion-dollar boondoggle." In 2012, the Canadian long-gun registry was finally killed and decommissioned by the then Conservative government.

HOMELAND SECURITY'S VIRTUAL FENCE

The US Department of Homeland Security was increasing the US Border Patrol's capabilities including radar, satellites, sensors, and communication links networks, often referred to as a "virtual fence." In September 2006, the Security Border Initiative (SBI) network contract awarded Boeing $20 million to establish a test portion along the Arizona–Mexico border for a stretch of 28 miles.

But early in 2009, Congress learned that the pilot project was delayed because users had been excluded from the process and the complexity of the project had been underestimated. In February 2008, the Government Accountability Office reported the radar could be confused by rain and other weather effects; the cameras produced images of uselessly low resolution beyond 3.1 miles, even though they were designed to be able to zoom in on subjects from much greater distances. In addition, local residents complained of interference with the Wi-Fi network communications system as a result of the pilot project. The project faced delays and cost overruns, and in April 2010, the SBInet project manager had resigned, citing the lack of a system design as only one of many concerns.

PROPER PLANNING

Can you see where most, if not all, of the disasters above could have been avoided? A good or even competent planning phase with the associated project analysis could have made all the difference.

For a contrast, we can look at the Spirit and Opportunity rovers. Those rovers were both designed and planned as part of the Mars Exploration Rover program. Launched in 2003, both Spirit and Opportunity landed successfully on Mars in 2004. The rovers were designed to gather data for a minimum of 90 days. Spirit landed on January 4, 2004 and finally failed on May 25, 2011. Its mission lasted more than seven years and the scientific data returned were also greater than ever expected. Opportunity has done even better and as of 2014 is still operational and gathering data.

The planning, analysis, and design work that went into this program were clearly unparalleled. The rover was to be rugged enough for the job, the contingencies were well planned for, and the testing obviously covered all the bases. This up-front work led to one of the most successful space exploration programs ever launched.

2

Academic History of Planning

Project—a planned undertaking.

THE STUDY OF PROJECT PLANNING

Traditional wisdom is that planning and analysis are important; the better the planning in a project, the more successful the project will be. Research as well as the experience of project and program managers has confirmed the importance of planning and that time spent on planning activities will reduce risk and increase project success. On the other hand, it is also thought that inadequate analysis and planning will lead to a failed project. If poor planning has led to failed projects, then perhaps trillions of dollars have been lost globally. This obviously has a major impact on projects and the economy as a whole.

But then we must ask: can there be too much planning? There is also a phenomenon in business called *analysis paralysis* (Rosenberg and Scott, 1999). This is when so much analysis takes place that no actual work is started, or it is started later than optimal. As well, "lightweight" project management techniques such as agile have been gaining popularity since they were first developed (Lindvall et al., 2002). Part of the ethos of agile methods is that less initial planning is better and an evolutionary process is more efficient. Agile methodologies even seem to imply that up-front planning is not effective. Agile advocates state that projects should not be planning too much up front; that may be too rigid to maximize customer benefit. Projects should evolve to best fit customer needs to be most successful.

The stakes are high as a company's success (and employees' futures) often depends on the success of its projects. Examples such as the Iridium satellite project (Collyer et al., 2010) and FoxMeyer exist where poorly planned and executed projects resulted in the bankruptcy of the companies.

I recall working on a project in the 1990s where a start-up sold a project to a major corporation with wildly optimistic goals. They claimed they could scan 100,000 documents per day and optically recognize handwriting with 80% accuracy. After everything was agreed to, I was brought in to manage the project to try to meet the goals. When those goals were clearly not going to be met, the contracts were very carefully reviewed and the lawyers brought in. If the project leadership had spent more time in up-front analysis, they might have had a much clearer understanding of the difficulty of the goals and the fact they were unlikely to be met. I did not survive as project manager, which I now consider positive. I moved on to another project and eventually, another company which no doubt saved me from many contentious meetings with lawyers and executives and the associated unwanted stress. In many cases, as a project manager, you may be brought in to a project that has already experienced failures during the planning phase and at that point, there may be little you can do.

HISTORY OF THE STUDY OF PLANNING IN GENERAL MANAGEMENT

The literature on management notes the importance of planning at least as far back as early in the last century with the work of Gantt (1910) and his famous Gantt charts and Gulick (1936) whose book was one of the first to define planning in business. Gantt charts (Figure 2.1) are the ubiquitous bar charts used to illustrate project schedules. They were first widely used during World War I but have become more and more widely used for project planning since the advent of computers. I was first introduced to Gantt charts in the early 1990s. I have rarely seen other types of project scheduling diagrams used in the IT, telecom, or financial sectors since. The government and military sectors do make more use of network diagrams and PERT, however.

Goetz stated in 1949, "Managerial Planning seeks to achieve a consistent, coordinated structure of operations focused on desired ends.

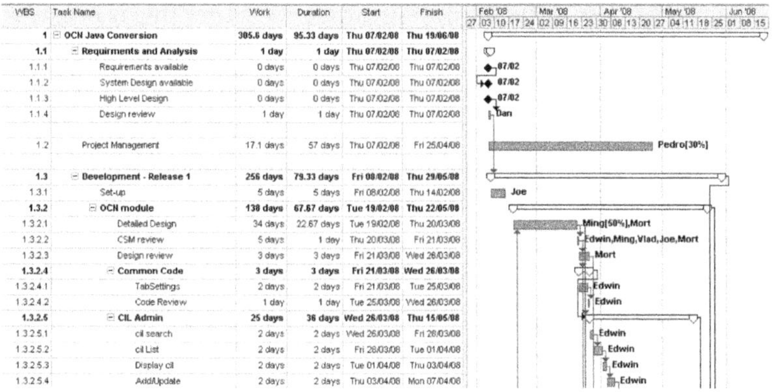

FIGURE 2.1
Sample Gantt chart.

Without plans, action must become merely random activity producing nothing but chaos" (p. 63).

That is not to say that everyone has always agreed with this assessment. From the 1970s on, some researchers such as Mintzberg and Kotter put greater importance on action and personal communication. Their views were that successful managers are able to react quickly to make key decisions and planning was not as important to running successful businesses as was flexibility. They stated successful leaders do not spend time locked away in detailed planning sessions; they spend their time meeting with key staff members, taking the pulse of the company, and reacting quickly to events. However, other authors have attempted to refute this view and made strong points that the importance of planning in management cannot be overlooked. Carroll and Gillen (1987), for example, found that planning is correlated to both group and individual success and is one of the most important of the management functions.

PROJECT PLANNING IN ACADEMIC LITERATURE

The study of planning and its specific relationship to project success started in the 1980s, as the study of project management became more widespread in the literature. During this time the software project management literature attempted to define ideal effort levels of planning. Subsequently, studies on this topic have not specified how much time to spend on each phase of the project life cycle. Whether this is because this guidance was found not to be effective, the diversity of technology projects greatly increased, or it simply fell out of favor is not clear.

Planning was one of the most consistently discussed topics in the early days of the project management research literature although it then started to decline as a topic. More has been written in the late 1990s and in the twenty-first century on project success. Planning has continued to be a key part of project management textbooks but even there, the level of detail in the planning sections has tended to diminish. Several papers have now found a strong link between planning phase completeness and project success so the study of planning has now again started to be more common in the academic project management literature. This fits the understanding that most project managers have about how important planning is to their job.

MODERN VIEWS ON PLANNING

It is clear the importance of planning is well entrenched in the project management community. Here is an example that illustrates that thinking. A nonacademic survey of project managers conducted on LinkedIn (see Figure 2.2) found that 64% of project managers considered project

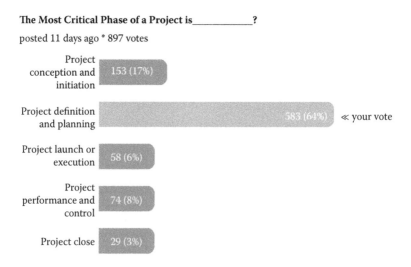

FIGURE 2.2
The most critical phase of the project survey from LinkedIn.

definition and planning the most important project phase, a much higher percentage than any other phase.

Although this survey is unscientific, it does hint at the experience and beliefs of practicing project managers regarding the key importance of planning. This general understanding of the importance of planning in project management seems to be ubiquitous among project managers. The business books available to project managers, however, have not taken up this trend, perhaps to the detriment of the project management field as a whole. Project management books nearly all devote time to planning and in some cases it is substantial. However, few, if any, base more than 60% of their content on planning as you could expect founded on the sentiments shown in Figure 2.2.

This could be because project management includes planning as one of its foundations. For example, planning is a key component of the *Guide to the Project Management Body of Knowledge (PMBOK® Guide)*—fifth edition (PMI®, 2013) and makes up the majority of the processes described there. This is also the case for other project management methodologies such as PRINCE2 (Murray, Bennett, and Bentley, 2009) from the British Office of Government Commerce (OGC). The question of whether planning is correlated with project success therefore may be a moot point. The benefits of planning have been confirmed through the practice of project management. It has thus become an expected part of all projects and project management.

ANALYSIS

An important part of the planning phase is the analysis and selection of options. In the *Oxford Dictionary of Philosophy* (Blackburn, 2005), *analysis* is defined as "the process of breaking a concept down into more simple parts, so that its logical structure is displayed." For proper planning, analysis is therefore a necessary part. If you think back to the failed projects mentioned in Chapter 1, a better planning phase could have avoided the problems encountered. But in most cases it was the analysis within those planning phases that should have uncovered the problems that ultimately led to project failure.

Analysis takes place throughout a project. However, it is particularly relevant in the planning phase. Requirements analysis and technical analysis are also usually required in projects during the planning phase. In addition, the project analysis effort can include efforts such as stakeholder analysis, requirements analysis, risk analysis, and systems analysis.

If we look at our failed projects of Chapter 1, stakeholder analysis was lacking in both the FoxMeyer and Gun Registration System examples. Key stakeholders were not identified and engaged (warehouse staff and affected agencies, respectively). The stakeholder impacts were major reasons for the project failures. Systems analysis and design issues were keys to the Mars Climate orbiter and Virtual Fence projects failures. In both cases, systems analysis failed to identify key technical risks (navigation units and camera performance). The architecture of the solutions also did not design in mitigation for these issues and failure was the result.

Researchers agree on the importance of analysis in project management. If we consider how Goetz in 1949 defined planning as "fundamentally choosing," analysis must occur during the planning phase. Informed choices can clearly not be made without analysis; analysis should precede and be a part of the planning process. One can even argue that the process of analyzing a project in detail as part of a planning exercise is one of the most beneficial aspects of doing planning in the first place.

PROJECT PLANNING PRACTICE

We next need to define more rigorously what is meant by planning. A classic definition of planning is "working out in broad outline the things that need to be done and the methods for doing them to accomplish the purpose set for the enterprise" (Gulick, 1936, p. 13). Koontz (1958) defines planning as "the conscious determination of courses of action designed to accomplish purposes" (p. 48). Mintzberg (1994) describes planning as the effort of formalizing decision-making activities through decomposition, articulation, and rationalization.

Goetz (1949) defines planning as "fundamentally choosing" (p. 2) and as mentioned above notes, "Without plans, action must become merely random activity producing nothing but chaos" (p. 63). As all these definitions imply, a key part of the planning process is analysis.

Planning is defined in the project management literature as well as in the general management literature. As noted above, planning is a key component of the *PMBOK® Guide* and makes up the majority of the processes described there. Another widely used project management methodology, PRINCE2 (Murray, Bennett, and Bentley, 2009), has a similar view on the importance of planning: "Planning is essential, regardless of the type or size of the project; it is not a trivial exercise but is vital to the success of the project" (p. 61). Another definition of planning in projects that was suggested to me by Professor Aaron Shenhar is all activities that come before execution.

It is important that we understand what the planning phase is and how it fits into the other phases of a project. The basic phases of a typical project, including the planning phase, were defined fairly early in the project management literature as in Pinto and Prescott (1988), and are shown in Figure 2.3. The preceding main structure has continued with different terminology, but the concepts and divisions remain the same.

The definition between what is initiation and what is planning may vary between sources and industries (Hamilton and Gibson, 1996; PMI, 2013; Kerzner, 2003). We assume the analysis tasks completed prior to execution to be part of planning. This is a broader definition of the planning phase than that of the *PMBOK® Guide* but is a closer match to definitions of planning in the construction industry (Hamilton and Gibson, 1996) and by Kerzner (2003). The most useful definition of the planning phase for

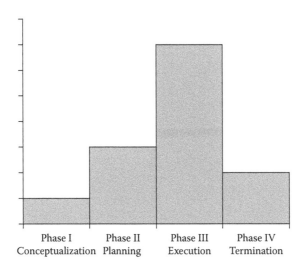

FIGURE 2.3

The stages in the project life cycle. (After Pinto and Prescott, *Journal of Management* 14(1): 5–18, 1988.)

the purposes of this book, however, gives the greatest flexibility and access to the widest range of data. To do otherwise would necessitate the removal of initiation- or conceptualization-related activities from discussion. For the purpose of this book, we define the planning phase as follows:

Planning Phase: The activities that come before execution phase in a project.

This definition does not differentiate preliminary planning and analysis that occurs before formal approval. However, it is assumed that this preliminary work is typically a small fraction of the overall planning. Hamilton and Gibson (1996) report this is typically only 5 to 10% of the overall design in construction projects, for example. The intention is to include this initial effort in the greater planning phase. The benefit of this approach is that it simplifies our analysis and discussion.

This definition of planning is not always consistent, however, and the methods of measuring planning vary. Some authors may measure the effort in money or person hours expended on planning activities. Others measure the quality and completeness of the planning deliverables. We therefore use the following definitions:

Planning effort: The amount of effort in money or work hours expended in planning.

Quality of planning: The quality or completeness of components of the planning phase or the quality of the planning phase overall.

3

Reasons Not to Plan

Life is what happens to you while you're busy making other plans.

Allen Saunders

Although the majority of the literature and books available highlight the importance of planning, that is not to say that there are no dissenting voices. You may also ask: "Is planning really as important as the traditional literature says? Is planning useful at all?"

VALUE IN PLANNING?

Anderson's 1996 paper titled, "Warning: Activity Planning Is Hazardous to Your Project's Health," speaks for itself. He questions the assumption that project planning is beneficial. He wonders if its benefits are real and asks, "How can it be that project planners are able to make a detailed project plan, when either activities cannot be foreseen or they depend on the outcomes of earlier activities?" (p. 89). Managers must make decisions early in the project, when they probably know little of the project's future direction. He questions the value of detailed planning from a conceptual standpoint.

Bart (1993) made the point that in research and development (R&D) projects, too much planning can lead to failure as formal control limits creativity, which is crucial in R&D. However, he also found that in some cases managers reduced control and planning too far, to the point that it was detrimental to the project. No planning and control clearly resulted in poor outcomes.

ERRORS IN PLANNING

Are there aspects of planning that make it prone to errors?

Buehler, Griffin, and Ross (1994) discuss the so-called "Planning Fallacy" where people consistently underestimate the time required to complete their own tasks while often overestimating the time required for others to complete their tasks. This phenomenon has been studied in psychology and found to be widespread and persistent even in the face of negative feedback on accuracy. People have a greater belief in their ability to complete work early than they can actually accomplish. Participants were asked how early prior to a deadline they would finish some upcoming projects in their lives. Even when shown their estimates were incorrect in previous rounds, they continued to be optimistic that they would finish the next set of work before their deadline, even though it had already been shown that was unlikely to be the case.

Brunnermeier, Papakonstantinou, and Parker (2008) describe the following planning fallacy pattern: "People initially underestimate the amount of work that the project will require and so, on average, do less than half the work in the first period. They gain the benefits of doing little work in the first period and expecting little work in the second. They lose some of the benefits of optimally smoothing effort and on average suffer in the second period when they have more than expected to do" (p. 7). Missed deadlines are often the result. They also note the phenomenon where people underestimate their own completion times but overestimate the times others will need to complete their tasks, particularly if the others are less experienced in that task. The planning fallacy literature highlights the need to consider these effects when planning.

We often tend to be overoptimistic when planning our lives. We think we will get that paper done a couple of days early so we can relax on the weekend. However, we end up working all day Sunday to make the deadline. We plan to get that bathroom renovation done over two weekends but it usually takes many more. If we don't have a thorough planning phase in our projects, we may not catch this inherent overoptimism.

PLANNING AND BUSINESS SUCCESS

But isn't planning critical for business success?

Orpen (1985) reported that small businesses which spent more time on strategic planning performed no better than other small businesses. However, the high-performing companies in his survey did perform more structured strategic planning and planning that was more thorough. This included having formal planning committees, having a higher opinion of the importance of planning, taking competitive companies into account in planning, and updating their plans on a regular basis; these tasks were linked to higher company performance. At the least, structured planning is important to company success. It appears companies that don't do strategic planning in a careful structured way are not as successful in the marketplace as those that do; and quality is better than quantity in this case.

PLANNING AND AVOIDING FAILURE

But with all the big projects that are delivered late and over budget, why didn't planning help?

They must have had a detailed planning phase. Flyvbjerg, Holm, and Buhl (2002) investigated 258 transportation infrastructure projects, worth US$90 billion and representing different project types, regions, and time frames. They found overwhelming statistical significance that the cost estimates used to decide whether such projects should be built are misleading. This highlights that even if planning and analysis are undertaken, senior management can choose not to use the resultant information whether or not it is good. However, van Marrewijk and colleagues (2008) state that large infrastructure projects are more often affected by political and structural problems than misleading estimates. They also state additional planning and control cannot resolve these issues. Where these types of issues do not exist, a project can be successful. It is therefore more important to be concerned about negative political or structural problems in mega projects then the possibility of misleading estimates. We probably all suspect that poor estimates are more likely to be a direct result of political decisions than a failure of a project team. Upper management may decide lowballing estimates is the best way to get the contract. Politicians may decide they can best sell the project with lower estimates and then worry about approving cost overruns later. In these cases the project team is just following directions while knowing they will be late and over budget.

> Shouldn't good planning be resulting in more successful projects as project management and project planning matures?

Magazinius and Feldt (2011) suggest that although estimation techniques have improved over time, success at meeting estimates has not. They state that factors such as the aggressive desire to reduce budgets distort estimates. This can result in adverse outcomes such as incorrect projects being selected or overruns costing more than the original rejected estimate. They even highlighted projects that might have been more beneficial to the organization if they had been passed over. They noted that the effectiveness of planning is often sabotaged by other goals. Love, Edwards, and Irani (2008) similarly note that unrealistic client demands are a factor that contributes to the creation of erroneous contract documentation in the construction field. They also note that one of the most common causes of severe deviations was attributable to deficiencies in the planning phase.

Of course, coming up with a successful plan can be hard when not all contingencies can be anticipated.

Collyer and colleagues (2010) describe examples of projects such as the Australian submarine project and the Iridium satellite project. These projects were deemed failures because the technology and environment changed so much during the course of the projects that the originally planned project outputs could not be successful. They state that in dynamic environments, projects need to cope with changes in technology during the course of the projects. If planning assumptions fail, unsuccessful projects can often result. They also make the statement, "While useful as a guide, excessive detail in the early stages of a project may be problematic and misleading in a dynamic environment" (p. 109). Koskela and Howell (2002) note that it is very difficult or perhaps impossible to maintain a completely up-to-date plan. Without an up-to-date plan, they believe that project work transforms to something similar to informal management.

McFarlan (1981), in reviewing three case studies in the IS field, notes both the benefits of planning and its limitations: "Formal planning and control tools give more subjective than concrete projections, and the great danger is that neither IS managers nor high level executives will recognize this. They may believe they have precise planning and close control when, in fact, they have neither" (p. 148). He showed how even effective planning cannot foresee certain complex technical failures. These failures are rare and costly and therefore planning for them is very difficult. In an example in one of the cases, a phantom screen appeared which required both software and hardware changes to fix at a cost of $200,000 in 1981 dollars. The vendor even had to create a new chip from scratch to resolve the problem. If these kinds of unexpected occurrences happen 1 in 100 times, how do you price in the cost? For 99% of the time the unexpected event will not occur and the budget will be too high. The other times, this cost would be very difficult to estimate. How could one predict it would cost $200,000 in the first place rather than $100,000 or $1 million? It may be impossible to plan for these types of occurrences or for all possibilities.

I had a similar personal experience. We were using a highly respected workflow tool, considered by many to be best in class. The case for the tool seemed good and the risk low. However, this was the first version built on a particular flavor of UNIX. That port of the product was rife with bugs that did not exist in other versions of UNIX and that were very hard to track down. In effect we were betting the project on version 1.0 of a tool. In the end it got so bad we had to abandon it and build our own code to replace it. Needless to say, we incurred a major delay and at points we thought we would need to abandon the entire project. Not a fun time for the project managers. Could we have planned for this? It's arguable if we could have.

Fitzgerald (1996) in reviewing software methodologies also notes "that later phases depend on the successful completion of earlier phases which requires perfect foresight" (p. 12). Later phases are much more likely to take place in the context of predecessors that did not run perfectly. And when asked about the key critical success factors (CSFs) for success most people do not tend to think of planning.

Poon and colleagues (2011), in using fuzzy set analysis on five case studies, found that high-level planning was ranked second lowest in importance out of five critical success factors studied. Planning was ranked below top management support, user involvement, and methodology.

The implication was that these CSFs were more important to project success than high-level planning.

Of course, in some environments planning is more difficult because of changing requirements. Collyer and Warren (2009) state that in dynamic environments events arise faster than they can be accommodated by replanning. They believed that creating detailed long-term plans in the first place for these projects can waste time and resources and lead to false expectations. They note that in dynamic environments, organizations tend to opt for high-level plans and for detailed planning they use techniques such as rolling wave, where planning is revised and redone at each phase. Aubry, Hobbs, and Thuillier (2008) in a study of project management offices (PMOs) note that for at least one organization studied, overly formal planning processes resulted in an impediment to the rapidity required to sustain successful projects. The three other organizations examined in the case studies did not report this effect but it apparently can happen. Kapsali (2011) reviewed the project management literature on innovation and concluded that in many papers too much planning was seen as detrimental to innovation. For example, she states that for innovation projects, "The uncertainty, complexity, and uniqueness of project activities make control more difficult and deviation from plans more probable, because plans are formulated for a set of contingencies that cannot be preconceived because they have no precedent" (p. 397).

Planning and its impact on success may also depend on what industry is involved. Zwikael and Globerson (2006) note that even though there is a high quality of planning in software and communications organizations, these projects still have relatively low ratings on success. They note this effect may be due to riskier technologies and environments, control issues, or overly ambitious commitments. Chatzoglou and Macaulay (1996) make the following points on why planning is sometimes shortened or eliminated in information technology (IT) projects:

The arguments/excuses of the project managers for not using a plan are:

- quality of the system is all that matters for a development process to be considered successful and the system useful,
- existing planning models give inaccurate and unreliable predictions and depend on many input variables that most of the time cannot be estimated in the beginning,

- and time is usually very limited and it is better to skip the planning and to start developing the requested system.

However, experience shows that none of the above arguments are valid. (p. 174)

Lack of planning is likely to lead to incorrect assumptions and poorly thought-out execution; the rework required to fix these mistakes will usually use up significant portions of project time, perhaps much more than what would have been spent in planning.

PLANNING AND DELAYS

And there may be further reasons why planning is not done as well as it could be. Chatzoglou and Macaulay (1996) make some points as to why it is rare that too much time is spent on requirements planning, or as they term it, requirements capture and analysis (RCA). "It is well known that the more time dedicated to the RCA stage of the systems development process, the fewer errors and mistakes will emerge later on in the next stages of the development process, and thus less time and costs will be needed for the development of the system. However, there are always some limitations that must not be exceeded because otherwise the overall result will be negative" (p. 175). Deadlines need to be met; any delay in the planning phase will result not only in the increased cost of the planning stage but also in a chain reaction in the next phases of the project.

Thomas et al. (2008) write, "Project managers are constantly pressured to 'get started with work' or 'make progress' by senior management who fail to recognize the value of planning in a project. These statements are widely supported by the comments received from the project managers in this study" (p. 109). This statement illustrates that in most projects there are pressures to reduce the time and effort spent on the planning phase. Brunnermeier et al. (2008) note that from a psychological standpoint, incentives for rapid completion of tasks and projects increase misplanning. Wideman (2000) notes that changing a detailed plan in construction projects can add cost and risk: "It is not generally understood that it is far more difficult, and costly in design time, to make major changes to an existing design than it is to start from scratch. This is because of the added careful coordination required and the higher probability and danger of overlooking the impact on a related system" (p. 9). Finally, Dvir, Tsvi, and Shenhar (2003) make this statement, "Although there are some that claim that too much planning can curtail the creativity of the project team, there is no argument that at least a minimum level of planning is required" (p. 89).

Few authors suggest no planning should be done or even that it is not important. The literature does not support the conclusion that planning should not be done in projects although certain caveats are highlighted.

4

Project Success and Planning: What Are They?

Success depends upon previous preparation, and without such preparation there is sure to be failure.

Confucius

Before it is possible to get into the detail of the impact of the project planning phase on success, it is of course important to define what a successful project is. Traditional measurements of project success focused on meeting the timelines and budget goals of a project. This is the so-called triple constraint or iron triangle. See Figure 4.1.

THE TRIPLE CONSTRAINT

This view of project success has become common and even ubiquitous. However, project success is also defined in a broader way. Although the measurement of project success has focused on tangibles in the past, current thinking is that, ultimately, project success can be better judged by the sponsors and stakeholders. After project completion, the impact on the customer and customer satisfaction becomes more relevant as do other aspects with a wider impact on the business. These aspects can include increased profits to the business, increased market share, greater efficiency, or faster time to market. Benefits can also be related to preparing organizational and technological infrastructure for the future, building skills and competencies in the organization, or opening new opportunities for markets, ideas, innovations, and products. This is not to say that the iron triangle is not important in project management.

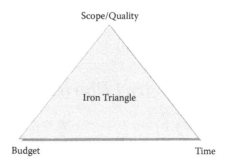

FIGURE 4.1
The triple constraint.

I have used it many times to illustrate to stakeholders that one cannot get something for nothing.

I once managed an imaging software project. The team had worked with me to build the plan and had fully bought into it. As the development completion date approached, the lead programmer realized that they would not make the deadline and would need to push back the start of testing. He decided that he would not let that happen. He embarked on a 48-hour coding marathon to complete the final parts of the program and did not go home for two days. Low and behold, the deadline was met, and everyone thanked him for his Herculean efforts. When testing started the following Monday, however, the problems with this strategy became apparent. The code did not work properly and completely failed the first set of tests. Upon review it was found that the code was not only buggy but it had major design flaws. Someone writing code at 4:00 a.m. is not going to make as well-thought–out design decisions as someone who has the time and energy to ponder and analyze his or her design. In the end, the testing had to be halted, and the code redesigned and rewritten. The marathon coding session probably had not helped and may have hindered progress. In this case, the project time was reduced, but quality definitely suffered; just as it should happen according to the iron triangle.

PROJECT TIME FRAMES

Much of the literature prior to 1980 considered that projects end when they are delivered. And that is when project management ends also. This is understandable from a project and project manager's standpoint.

The definitions of a project imply an end date; at that time the project manager is likely to be released or move on to another project. However, the study of project management has also at times examined the wider impact of projects.

Pinto and Slevin (1988) stated, "There are few topics in the field of project management that are so frequently discussed and yet so rarely agreed upon as the notion of project success" (p. 67). Shenhar, Levy, and Dvir (1997) note that of the three traditional dimensions of project efficiency—time, budget, and scope—scope can have the largest role. Not only is scope an aspect of project efficiency, but it also has an impact on customers and their satisfaction. They note "Similarly, project managers must be mindful to the business aspects of their company. They can no longer avoid looking at the big picture and just concentrate on getting the job done. They must understand the business environment and view their project as part of the company's struggle for competitive advantage, revenues, and profit" (p. 10). They also stated that a low-tech project is more likely to rely on efficiency as a measure of success, for example.

BROADER SUCCESS

Jugdev and Müller (2005) reviewed the project success literature over the past 40 years and found that a more holistic approach to measuring success was becoming more in evidence. Researchers were increasingly measuring success by the impacts on the organization rather than success at only meeting the triple constraint. They noted that a project can be a success despite a poor project management performance. The movie, *Titanic*, was touted as a late overbudget flop but went on to be the first film to generate more than $1 billion in revenue.

The Sydney Opera House was years late, and many times overbudget. How late and overbudget was it? The original cost estimate was set at $7 million in 1957. The original target completion date was January 26, 1963 (Australia Day). The Opera House was finally completed in 1973; the final budget was $102 million. The project was delivered 10 years late and ran overbudget by more than 14 times but it has gone on to become the symbol of Sydney. As one of the most famous buildings in the world, it is hard to define it as a failure. And most have forgotten how late and overbudget it was.

The importance of broader success measures for projects is now the norm. The most recent version of the *PMBOK® Guide* as an example, no longer mentions the triple constraint (*PMBOK® Guide*, PMI®, 2013). It now includes customer satisfaction in addition to time, budget, and scope. One can argue, however, that time, budget, and scope goals are still an important part of project management.

EFFICIENCY AND SUCCESS

Müller and Turner (2007) defined 10 dimensions of project success as part of their series of studies on project manager competencies and project success. Meeting time, budget, and scope goals was the dimension of success mentioned most often by participants, twice as often as any of the other measures in that study. It is clear that aspects of efficiency still have importance for project managers.

Zwikael and Globerson (2006), using data collected from 280 project managers, showed that aspects of success demonstrate a similar frequency distribution. Figure 4.2 shows a highly similar distribution between technical performance (project efficiency) and stakeholder satisfaction.

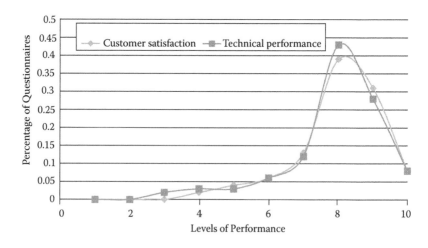

FIGURE 4.2

Frequency distribution of technical performance and customer satisfaction. (After O. Zwikael and S. Globerson, *Benchmarking: An International Journal* 13(6): 688–700, 2006. With permission.)

Here technical performance was analogous to meeting scope requirements. In addition, they reported a linear correlation between technical performance and customer satisfaction. This result showed a strong relationship between the two.

As well, Dvir, Tsvi, and Shenhar (2003) also found that "all four success-measures (Meeting planning goals; End-user benefits; Contractor benefits; and Overall project success) are highly inter-correlated, implying that projects perceived to be successful are successful for all their stakeholders" (p. 94). Quality or scope is closely related to issues of specifications, technical performance, and functional objectives. This will clearly affect the perception of multiple project stakeholders. Other researchers have stated that all measures of project success contain the traditional success factors of time, cost, and performance. In the light of some of these studies, it is possible that PMI's decision to remove the triple constraint from the most recent version of *PMBOK® Guide* was premature.

It may be that projects that are well managed from a project efficiency standpoint may also be better run from an overall success standpoint, and the papers mentioned above may point to that being the case. Therefore for the purpose of this book, I do not ignore either project efficiency or overall project success but discuss both aspects of project success.

Therefore I refer to

Project efficiency: Meeting cost, time, and scope goals; and
Project success: Meeting wider business and enterprise goals.

With most studies of project success that use questionnaires or interviews, the results rely on participants stating how successful a project was. This is subjective by nature. One could argue that there may be ways to measure success in an objective way; however, this likely only applies to project efficiency. Therefore this book is largely concerned with perceived project success as we cannot pretend to be fully objective in the measurement of success. As has been stated many times, perception is often reality.

And as previously discussed, for typical projects, success and project efficiency are often correlated (Dvir et al., 2003; Zwikael and Globerson, 2006). However, are there ways we can measure this correlation? I discuss this in the next chapter.

5

Efficiency versus Success

Eighty percent of success is showing up.

Woody Allen

Project success has long been a topic of interest in the project management literature. One of the greatest directions in the study of success over the past 20 years is that time, budget, and scope goals, so-called project efficiency, is not the best way to measure success: broader success measures should be used. However, is efficiency still important and to what extent? The relationship between efficiency and broader success has not been well studied empirically and this book investigates this relationship.

In my research, I collected data on more than 1,300 projects from more than 60 countries. I use those data for the remainder of the analyses for this book. Details on how the data were collected and analyzed can be found in Appendix A.

When I designed my study, I took the view that a relationship could be found between effort in the project planning phase and perceived project success. Because perception and observation are at least partially based on subjective opinion, my results cannot be fully objective. Some concepts such as project success may not be fully quantifiable and are affected by subjective judgment of the participants and sponsors. This book deals with quantities such as effort or cost of planning phase, effort or cost of the overall project, and percentage of project effort that was dedicated to the planning phase.

To facilitate this analysis, the success questions were grouped in three success measures. These are the measures of project success used throughout this book. They are as follows:

Efficiency Measure = mean of the following three responses as a summated scale:
1. How did the project do in meeting project budget goals?
2. How did the project do in meeting project time goals?
3. How did the project do in meeting project scope and requirements goals?

Stakeholder Success Measure = mean of the following four responses as a summated scale:
1. How did the project sponsors and stakeholders rate the success of the project?
2. How do you rate the project team's satisfaction with the project?
3. How do you rate the client's satisfaction with the project's results?
4. How do you rate the end users' satisfaction with the project's results?

Success Measure = mean of the following eight responses as a summated scale:
1. How do you rank the overall success of the project?
2. How did the project do in meeting project budget goals?
3. How did the project do in meeting project time goals?
4. How did the project do in meeting project scope and requirements goals?
5. How did the project sponsors and stakeholders rate the success of the project?
6. How do you rate the project team's satisfaction with the project?
7. How do you rate the client's satisfaction with the project's results?
8. How do you rate the end users' satisfaction with the project's results?

These three measures along with the question put to the respondents to rate the overall success of the project was used for the analysis. These measures were analyzed for consistency (see Appendix B) and found to be quite consistent within all the measures. Scope was found to be the efficiency component most closely related to the overall success measures which is in keeping with Shenhar, Levy, and Dvir (1997) who stated that scope was the most important component of the triple constraint for overall success.

Traditional measurements of project success focused on meeting the timelines and budget goals of a project (Kerzner, 2003). Munns and Bjeirmi (1996) noted that much literature to that point considered "projects end when they are delivered to the customer. That is the point at which project management ends. They do not consider the wider criteria which will affect the project once in use" (p. 83). This is understandable. The definitions of a project imply an end date; at that time the project manager is likely to be released or move on to another project.

From the papers reviewed in Chapter 4, we see that project efficiency and success are not the same thing but they must be somehow related.

To understand the nature of the relationship, I first studied the correlation between the respondents' project success rating and the success measures. You may recall from your past statistics courses that a correlation is the degree to which two or more measurements on the same group of elements show a tendency to vary together. A correlation of 1 shows the elements are probably measuring the same thing; a correlation of 0 shows there is no relationship at all between them.

This success rating was one question that asked them to rate the overall success of the project. The other measures were combinations of a number of questions.

The analysis shows close to 90% correlation between this one question and the main success measures other than the efficiency measure. This indicates a very close correlation between the managers' overall rating of project success and measures combining the wider success measures. However, the efficiency measure only shows a .58 correlation with the manager's assessment. See Table 5.1.

Another statistical method we should discuss is the *p*-value. The *p*-value is the probability of obtaining a test statistic at least as extreme as the one that was actually observed. In other words, it is the probability that that

TABLE 5.1

Correlation between Project Success Rating and the Success Measures

Correlations		
Marked Correlations Are Significant at $p < .050$ $N = 1,386$ Projects		
	Efficiency Measure	**Stakeholder Success Measure**	**Success Measure**
---	---	---	---
Respondents' Overall Project Success Rating	0.58[a]	0.87[a]	0.88[a]

[a] $p < .05$.

observed result was a fluke. A *p*-value of .1 indicates a 10% chance the result was random and not due to a true relationship. As is standard for most studies in economics, I use $p < .05$ as my cutoff for results in most cases. My results in general were much better than that cutoff. Note that the *p*-value is not the only important value in my analysis. A result can have a very good *p*-value but with such a low R^2 value that the relationship is so weak as to be not important. I test for both factors.

PROJECT EFFICIENCY VERSUS PROJECT SUCCESS

When we compare correlations among all the measures and the project success rating, we see the results in Table 5.2.

The respondents' own overall success measure, project success rating, had a correlation of .87 with overall success and .88 with stakeholder success. The success measure that had the lowest correlation with all the other success measures was the efficiency measure, which had a correlation of .60 with the stakeholder success measure and .58 with the respondents' self-reported overall success rating.

Table 5.3 shows the correlation of project success with the individual components of efficiency: time, cost, and scope. The correlation with the project success measure and the stakeholder satisfaction measure is between 0.4 and 0.6. The highest correlation with success is with meeting scope, as we would expect from the work of some researchers (such as Shenhar et al., 1997) who found that scope had the highest impact on project success of the efficiency measures.

TABLE 5.2

Correlations between Project Success Measures

	Correlations Marked Correlations Are Significant at $p < .050$ $N = 1,386$ Projects			
	Project Success Rating	**Efficiency Measure**	**Stakeholder Success Measure**	**Success Measure**
Project Success Rating	1.00	0.58[a]	0.87[a]	0.88[a]
Efficiency Measure		1.00	0.60[a]	0.83[a]
Stakeholder Success Measure			1.00	0.94[a]
Success Measure				1.00

[a] $p < .05$.

TABLE 5.3

Correlation of Individual Efficiency Measures to Project Success Measures

	Correlations Marked Correlations Are Significant at $p < .001$ N = 1,386 Projects			
	Project Success Rating	Efficiency Measure	Stakeholder Success Measure	Success Measure
Project Budget Goals	0.41[a]	0.83[a]	0.42[a]	0.63[a]
Project Time Goals	0.51[a]	0.88[a]	0.51[a]	0.72[a]
Scope and Requirements Goals	0.54[a]	0.77[a]	0.58[a]	0.72[a]

[a] $p < 0.001$

Finally, I completed a regression analysis of the efficiency measure versus the success measure. Now we dive a bit more into statistics. I apologize as I hate statistics perhaps as much as you do!

The coefficient of determination R^2 provides a measure of how well future outcomes are likely to be predicted by a model. For example, assume you are shopping and have no credit cards, just $100 cash. The R^2 of the relationship between the amount of cash you have and maximum you can spend shopping is therefore 1.00. It is directly related to the amount of cash you have with you; the more you have the more you can spend. However, if you also have a credit card with a $400 limit on you, then the R^2 between your cash and the maximum you can spend is .20 ($100/$500). But because we are referring to social sciences and economics, it is not that simple. If we look at how much you will actually spend, the relationship will also need to factor in how good the merchandise or prices are at the store, whether you need money for dinner later, and how long until your next paycheck. Maybe the R^2 between the cash in your pocket and how much you will actually spend is much smaller, maybe .10 or .05. Therefore the amount of cash in your pocket may only be a 10% predictor of how much money you will spend shopping. Maybe it is less. To summarize, R^2 indicates how much one variable affects another. In social sciences $R^2 > .6$ indicates two variables are probably measuring the same thing. R^2 above .05 is significant although R^2 less than .05 can also be significant in certain circumstances.

Now, we're ready in Table 5.4 to examine R^2 with respect to how efficiency affects overall success. This analysis also shows a relationship between efficiency and success. It indicates with a quite low p-value that the efficiency measure is related to the success measure with an R^2 of .36.

TABLE 5.4

Regression Analysis for Efficiency Measure versus the Success Measure

Regression Summary for Planning Measure versus Success Measure				
	Number of Projects	R	R^2	p-Level
Planning Measure	1,386	0.602	.362	0.000

This could indicate that meeting a project's time, budget, and scope goals is 36% responsible for a project's broader success ratings.

As has been reported in the literature (Jugdev and Müller, 2005; Cooke-Davies, 2002; Shenhar et al., 1997) and reiterated in this research, project success and project efficiency are not the same thing. From my research, I can confirm that project efficiency is not the final measure of over-all project success although it is also not completely independent of it. Project efficiency was only correlated at roughly .6 with broader success measures, whereas the broader success measures were highly correlated with each other.

PRACTICAL APPLICATIONS

What does this mean from a practical standpoint? Project success should be measured and planned to include success measures broader than the traditional time, cost, and scope measures. The traditional project success measure of project efficiency is only 60% correlated with overall project success after all. Project managers should be aware that when they plan project success goals that the broader success measures need to be taken into account and made part of the planning process.

However, the converse is also true. If you ignore efficiency, particularly scope, you are unlikely to deliver a project that will be viewed as fully suc-cessful, even in the long term. To paraphrase Woody Allen, the project has to show up (or be delivered) to the stakeholders for it to be considered successful at all. If it shows up on time and on budget, that will probably help and not hinder the long-term view of its success.

6

Planning Variation by Industry

The best laid schemes o' Mice an' Men,
Gang aft agley.
An' lea'e us nought but grief an' pain,
For promis'd joy!

Robert Burns

Different industries may require different types of projects and have different project management needs. This undoubtedly has an impact on the need for planning and the effect of planning on success.

Zwikael and Globerson (2006) found that construction and engineering had the highest quality of planning and highest reported success, whereas production and maintenance companies had the lowest quality of planning and lowest reported rates of success. "Construction and engineering organizations, which scored the highest on project success, also obtained the highest score on quality of planning. Production and maintenance organizations, which scored the lowest on project success, received the lowest quality score as well" (p. 694). They also noted this is not the case for software and communications organizations. They had a high degree of planning but still often delivered projects with poor results. They stated that this may be due to riskier technology and environment, control issues, or overly ambitious projects. There is, of course no way to know if projects in these industries would have been even less successful had a high quality of planning not been in place.

The production and maintenance industry is less project focused and may have less of an entrenched project culture and less of an entrenched project planning culture. The services industry is third in planning and second in success, and software and communications were second in planning and third in success. These last two results can be attributed to either differences in the impact of planning on each industry or the fact software and communications industries may be challenging environments. I think it is clearly the latter. Project management training is popular in software and communications fields specifically because they are environments where failed projects are common. Good planning is required in this industry but may not fully counteract the challenges and complexities found there.

Zwikael (2009) also identified the importance of the *PMBOK® Guide's* nine knowledge areas to project success and analyzed the impact by industry. He showed that there is a marked difference in the types of knowledge areas that affect project success by industry. The study implies that the importance of planning, which areas of planning are most important, and the optimum quantity of planning can all vary by industry.

Early in my career, I worked as an engineer at a nuclear plant. It was there I started to understand the benefits of planning ahead, both professionally and personally. As a maintenance engineer, I was required to ensure that spare parts were available for the equipment. For less frequently used parts, this involved watching the inventory lists and ordering parts when there were one or two parts left in inventory. As you can imagine needing a part and then waiting a couple of weeks for delivery was not acceptable. A plant could not sit idle, producing no electricity for want of a single part, perhaps as simple as a small seal. So parts would always be ordered as soon as inventory went down to one or two items. I realized that applying this to life in general could reduce wasted time and hassles. Don't plan to buy supplies when you are out; write it down when you are almost out. Keep a running list on the fridge. I found with a little preplanning, emergency runs to the grocery store for milk, sugar, or toilet paper disappeared, and I could use that time for other things. There is a more general lesson here. Spending a bit of time earlier can save a lot of wasted time later.

Collyer et al. (2010) found that approaches to planning varied greatly within industries. They reported differences in formality of planning that varied with the dynamism of the environment. This ranged from less dynamic (construction and defense) to highly dynamic (film, venture capital, and technology). For example, a construction project will have highly detailed requirements (architectural blueprints) that should not be deviated from during the course of the project without careful change control. Similarly, a defense project will define the exact functions of the equipment being built. Film or venture capital, on the other hand, must be more flexible. If something is not working during execution, it will be quickly changed to assure success. For example, if a movie shoot is not going well, it is better to bring in new writers to do a rewrite than risk a flop. Or in venture capital, if a product finds an unexpected new market, managers need to rework the product and launch into that market. These environments are dynamic; the ability to react to change and to take advantage of unexpected opportunities is critical for these industries.

APPROACHES IN DIFFERENT INDUSTRIES

Collyer et al. (2010) also discuss some methods to overcome the challenges of delivering projects in dynamic environments (pp. 113–117):

- Make-Static Approach: Resist all change to original scope and plans. This is common in the construction and defense industries.
- Emergent Planning Approach: Plan from a high level initially and then complete detailed planning as you go. This was supported by many of the interviewees from various industries.
- Staged Releases Approach—Scope Reduction: Deliver many small projects rather than one large project. This has been used in the pharmaceutical and technology fields and may be a partial origin of the agile methodologies.
- Competing Experiments Approach: Work on more than one design or approach in parallel and select the one that best meets changing environments. This is used in the film, venture capital, and high technology fields.
- Alternate Controls Approach: Ensure a highly functioning team that reacts to a dynamic environment. This is done by careful team selection or by high rewards upon final project success.

Nobelius and Trygg (2002) noted that components of front-end activities vary between project types. For example, business analysis was found to be the number two priority for a project to build on an existing product line but was not found to be important in either a research/investigational project or in an incremental change project to an existing product. This is not surprising as different projects will require different staff and different requirements. Construction projects do not require business analysis, although for a project at a bank it might be one of the most important aspects of the planning phase. Given this variability among projects, is there anything we can say about the importance of planning to projects in general? I answer that in future chapters.

There are no definitive studies that link planning requirements to industries; however, the research does indicate that planning requirements vary between companies, from project to project, and that different industries require different planning and planning tools. Based on the research available, it is clear that planning varies.

Planning requirements vary in different industries and between projects.

Two industries have a more extensive body of research on planning and success: construction and information technology. For this reason they are given special consideration in Chapters 8 and 9.

7

Geography and Industry and Success

Let our advance worrying become advance thinking and planning.

Winston Churchill

NEW RESEARCH

One question of interest is whether project success varies in general between countries and industries. This area has been discussed in the literature but not thoroughly studied. Because the responses to my research were global in nature, it was possible to get a glimpse of how success can vary by location and by industry. As you may recall from Chapter 5 the success questions were grouped in three success measures. These are the measures of project success used throughout this analysis. They are as follows:

Efficiency Measure
Stakeholder Success Measure
Success Measure

These three measures along with the question put to the respondents to rate the overall success of the project were used for the analysis.

INDUSTRY

First the success measures were compared based on industry (Table 7.1). In addition to the three calculated success measures above, the respondents' single question, "Project Success Rating," was examined.

TABLE 7.1

Descriptives by Industry

	Success Measure	Efficiency Measure	Stakeholder Success Measure	Project Success Rating	Valid N
Construction	3.5	4.6	3.7	3.5	41
Financial services	3.3	4.6	3.4	3.4	257
Utilities	3.3	4.5	3.6	3.5	42
Government	3.4	4.7	3.4	3.4	152
Education	3.4	5.1	3.5	3.5	42
Other	3.3	4.5	3.2	3.2	157
High technology	3.4	4.8	3.5	3.5	223
Telecommunications	3.4	4.8	3.5	3.4	133
Manufacturing	3.2	4.3	3.3	3.3	122
Health care	3.4	4.9	3.4	3.3	113
Professional services	3.3	4.8	3.3	3.4	69
Retail	3.2	4.4	3.0	2.9	35
All Groups	3.3	4.7	3.4	3.4	1,386
$p(F)$	0.69	0.40	0.50	0.88	

We can see, for example, that construction has the highest project success measure. This is in agreement with the literature review results that the construction industry in general had better rates of success than other industries studied (Zwikael and Globerson, 2006). Retail and manufacturing report the lowest success ratings. Project management is well established in construction but probably not so well established in manufacturing and retail. Maybe this is not a coincidence.

However, other trends are more difficult to see and the ANOVA analysis ($p(F)$ indicating statistical significance) does not indicate any of the measures are significantly related to industry. It is interesting to note that some industries show higher success rates in either project efficiency or stakeholder success. Professional services, for example, shows a higher than average efficiency rating but lower than average project stakeholder success rating. This may not be that surprising inasmuch as professional services companies incur penalties if they deliver late or are missing scope items. However, longer term stakeholder satisfaction measures may not be written into their contracts.

In my own career, I have been involved in many projects involving consulting firms providing professional services. In planning for projects involving consulting firms, it is good to keep in mind that these firms need to make a profit. I have found that to get a contract they will send their most senior and experienced people to design the solution. However, the team that actually does the work will not be as senior and are sometimes best described as bright juniors. Junior staff doesn't demand as high a salary. Learning curves and mistakes that are due to inexperience are to be expected and if not planned for, can have a major impact on the success of a project. Firms may pull out all the stops to meet contractually defined timelines and budgets. However, other aspects may fall by the wayside. The results of my research seem to bear that out.

GEOGRAPHY

I now looked at variations by region, shown in Table 7.2. It is interesting to note that in this case, success ratings appear to vary by region. The ANOVA results show a significant *p*-value for all of the measures examined.

The Pacific reports the highest average efficiency followed by North America and then the Far East. Although Russia reports the lowest efficiency followed by Latin America and Africa, one can speculate that

TABLE 7.2

Descriptives by Project Location

	Success Factor	Efficiency Factor	Stakeholder Success Factor	Project Success Rating	Valid *N*
Indian subcontinent	3.32	4.51	3.39	3.38	97
North America	3.44	4.79	3.47	3.42	756
Africa sub-Sahara	3.20	4.44	3.24	3.11	37
Australasia	3.22	4.45	3.22	3.35	49
Europe	3.26	4.52	3.29	3.24	213
Latin America	3.07	4.23	3.10	3.10	83
Russia and FSU	3.25	4.17	3.42	3.42	12
Pacific	3.39	4.81	3.37	3.38	24
Middle East	3.25	4.43	3.32	3.23	82
Far East	3.14	4.77	3.00	2.88	32
All Groups	3.39	4.65	3.44	3.41	1,386
p(F)	*0.01*	*0.01*	*0.02*	*0.05*	

Africa is a more challenging environment and therefore has fewer projects coming in on time and on budget. On the other hand, North America reports the highest stakeholder success followed by Russia and the Indian subcontinent. The lowest stakeholder success was reported by projects in the Far East, Latin America, and Australasia.

Stakeholder success and efficiency do not seem to be linked in the rankings of success by regions. Areas that report better efficiency do not always report greater overall success.

Next, we review the means for scope of projects: local versus international as shown In Table 7.3. There do not appear to be strong differences between the means for this measure although national projects appear to have the highest reported success rates. The ANOVA results confirm this with no significant p-values found, although the p for efficiency is close to the threshold at $<.10$.

We can also look at how efficiency and the other success measures are correlated by industry, Table 7.4. That is, how important are meeting time, budget, and scope goals to the overall success rating of the project. It is interesting to note that efficiency is most highly correlated by stakeholder success in utilities, healthcare, and professional services. This correlation implies that meeting time, budget, and scope goals affect other broader success measures. It is least correlated for government and high technology. This result for high technology may surprise some people, although the result for government might not surprise many.

We can look deeper into these relationships by breaking down which components of efficiency were most important by industry. Kerzner (2009, p. 736) breaks down which industries are most likely to sacrifice time, cost, or scope (performance) when trade-offs are required. We can use that as a basis for comparison. Table 7.5 shows that budget goals and

TABLE 7.3

Descriptives by Local versus International Projects

	Success Measure	Efficiency Measure	Stakeholder Success Measure	Project Success Rating	Valid N
One city or region	3.3	4.7	3.4	3.3	577
National	3.4	4.7	3.4	3.4	367
International	3.3	4.5	3.4	3.4	442
All Groups	3.4	4.6	3.4	3.3	1,386
p(F)	*0.44*	*0.09*	*0.70*	*0.61*	

TABLE 7.4

Correlation of Efficiency versus Other Success Measures by Industry

	Overall Project Success Rating	Stakeholder Satisfaction Measure	Overall Success Measure	Valid N
Construction	0.530	0.635	0.845	41
Financial services	0.635	0.680	0.859	257
Utilities	0.744	0.706	0.896	42
Government	0.465	0.410	0.730	152
Education	0.592	0.627	0.852	42
Other	0.507	0.579	0.815	157
High technology	0.498	0.515	0.806	223
Telecommunications	0.664	0.651	0.860	133
Manufacturing	0.692	0.687	0.878	122
Healthcare	0.606	0.694	0.868	113
Professional services	0.658	0.673	0.854	69
Retail	0.598	0.616	0.830	35

Note: All results above were significant at $p < .001$.

TABLE 7.5

Stakeholder Success Measures versus Efficiency Components by Industry

	Budget Goals	Time Goals	Scope Goals	Valid N
Construction	0.47	0.71	0.44	41
Financial services	0.50	0.57	0.66	257
Utilities	0.55	0.55	0.70	42
Government	0.30	0.30	0.42	152
Education	0.094[a]	0.57	0.70	42
Other	0.41	0.47	0.57	157
High technology	0.39	0.43	0.46	223
Telecommunications	0.40	0.56	0.68	133
Manufacturing	0.49	0.67	0.56	122
Healthcare	0.46	0.57	0.61	113
Professional services	0.45	0.56	0.66	69
Retail	0.35	0.47	0.70	35

[a] Marked result was not significant at $p < .05$. All others were significant.

stakeholder success were most correlated for utilities, financial services, and healthcare. Interestingly, this is in full agreement with Kerzner (2009) for utilities and healthcare although not for financial services. Table 7.5 paints an interesting picture of how priorities may vary between industries. This analysis is in general agreement with Kerzner (2009); approximately half of the findings are in agreement and those remaining are in partial agreement or in contradiction.

Therefore we can conclude from the overall evidence and as one would expect from the previous research

---

Success rates vary between regions and industries.

8

Planning in the Construction Industry

A goal without a plan is just a wish.

Antoine de Saint-Exupéry

Project management has a long history in the construction industry and there have been a number of studies in the construction project management field on the relationship between planning and project success: this is a well-studied area in comparison to other industries or other areas in project management. The construction industry was key to the development of project management as a discipline and still features prominently in the project management literature.

RESEARCH IN CONSTRUCTION

Faniran, Oluwoye, and Lenard (1998) in a survey of Australian construction firms found that "Planning time was identified as the critical determining factor for the extent to which the actual construction cost varied from the value originally stipulated in the contract documents at the time of award of the contract" (p. 252). They reported that planning time explained 16% of the variance in cost. Similarly, the proportion of planning time spent gathering information or requirements explained 14% of the variation in quality and together with past construction experience, 18% of the time variance. They reported a correlation of planning time to project cost: the project total budget was correlated with increased time spent in planning. This is to be expected as more planning should be undertaken for larger, more expensive projects. Only time spent on planning was studied here, however, and not effort or budget. Finally, they report that construction planning effectiveness can be increased by

1. Investing substantial quality time in planning prior to the commencement of work onsite
2. Focusing on choosing the appropriate construction methods rather than placing the focus on the development of schedules

This is one of a few papers to study planning time and its link to success. However, the analysis focuses on planning's impact to variances and not planning's impact on project success.

Love, Edwards, and Irani (2008) studied findings on the effects of errors during the design portions of the planning phase on rework. They noted that "Design-induced rework has been reported to contribute more than 70% of the total amount of rework experienced in construction and engineering projects" (p. 234). In the case studied, they found that the errors in design resulted in cost overruns, which consumed all the profit for the participating firms. Although the customer in this case received the building ahead of schedule, the construction firms involved considered the project a failure. One can imagine this is the case for many failed projects: errors during planning resulting in project failure. Chapter 1 illustrated a number of examples from the Mars Climate Orbiter to the Canadian gun registry.

Hamilton and Gibson (1996) found that an increase in preproject planning quality for construction projects increased the likelihood of a project meeting financial goals. The top third of projects from a planning perspective had a much better chance of meeting those goals, whereas fewer of the projects in the lower third of projects did. Similar results are seen in these projects' results relating to schedule performance and design goals met.

Shehu and Akintoye (2009) found in a study of program management in the UK construction industry that effective planning was the number one critical success factor (CSF) identified. Effective planning had the highest criticality index of the 22 CSFs identified. In this study it was stated that program management involves management of a group of projects and that CSFs for project management and program management may differ. However, they also note that "the relationships project management and programme management are observed to be synergistic" (p. 2). This book is focused on project management rather than portfolio or program management. However, it is interesting to note that planning is seen as important for portfolio and program success as well as project success. Planning is important for programs and project portfolios; however, detailed planning is typically completed at the project level. Studying how best to plan in the program and portfolio environment is a topic for a future book.

PDRI

A method of measuring the quality of project planning has come into use in the construction industry. The Project Definition Rating Index (PDRI) is a method to measure project scope definition for completeness. Developed by the Construction Industry Institute (CII) in 1996, this tool has been widely adopted by various owners and designers in the building industry. It has gained acceptance in the facilities and construction industry as a measure of the quality of preproject planning. The PDRI offers a comprehensive checklist of 64 scope definition elements in a score sheet format. A PDRI score of 200 or less has been shown to increase greatly the probability of a successful project. Undertaking no planning correlates to a PDRI score of 1,000. Table 8.1 gives a detailed breakdown of the components of the PDRI.

Gibson et al. (2006) noted that research results show that effective preproject planning on industrial and building projects leads to improved performance in terms of cost, schedule, and operational characteristics.

This study also found that "Project performance and PDRI data were collected from the sample projects and showed that the PDRI score and project success were statistically related; that is, a low PDRI score (representing a better-defined project scope definition package just prior to detailed design) correlates to an increased probability for project success" (p. 37).

In addition, Gibson and Pappas (2003) reported the results shown in Table 8.2, demonstrating a marked difference in empirical measurements of project success based on the project PDRI score. The table looks at the impact of the PDRI scores on the success of building projects.

Table 8.3 looks at the impact of the PDRI scores on the success of industrial projects. The average overall for all projects was a PDRI score of 238.

Furthermore, they note, "Indeed, due to the iterative and often chaotic nature of facilities planning, many owners face such uncertainty that they skip the entire planning process and move to project execution, or decide to delegate the pre-project planning process entirely to contractors, often with disastrous results" (p. 41).

TABLE 8.1

PDRI Sections, Categories, and Elements

SECTION I. BASIS OF PROJECT DECISION

A. Manufacturing Objectives Criteria
A1. Reliability Philosophy
A2. Maintenance Philosophy
A3. Operating Philosophy

B. Business Objectives
B1. Products
B2. Market Strategy
B3. Project Strategy
B4. Affordability/Feasibility
B5. Capacities
B6. Future Expansion Considerations
B7. Expected Project Lifecycle
B8. Social Issues
B1. Products

C. Basic Data Research and Development
C1. Technology
C2. Processes

D. Project Scope
D1. Project Objectives Statement
D2. Project Design Criteria
D3. Site Characteristics Available versus Required
D4. Dismantling and Demolition Requirements
D5. Lead/Discipline Scope of Work
D6. Project Schedule

E. Value Engineering
E1. Process Simplification
E2. Design and Material Alternatives Considered/Rejected
E3. Design for Constructability Analysis

II. FRONT-END DEFINITION

F. Site Information
F1. Site Location
F2. Surveys and Soil Tests
F3. Environmental Assessment
F4. Permit Requirements
F5. Utility Sources with Supply Conditions
F6. Fire Protection and Safety Considerations

G. Process/Mechanical
G1. Process Flow Sheets
G2. Heat and Material Balances
G3. Piping and Instrumentation Diagrams (P&IDs)
G4. Process Safety Management (PSM)
G5. Utility Flow Diagrams
G6. Specifications
G7. Piping System Requirements
G8. Plot Plan
G9. Mechanical Equipment List
G10. Line List
G11. Tie-in List
G12. Piping Specialty Items List
G13. Instrument Index

H. Equipment Scope
H1. Equipment Status
H2. Equipment Location Drawings
H3. Equipment Utility Requirements

I. Civil, Structural, and Architectural
I1. Civil/Structural Requirements
I2. Architectural Requirements

J. Infrastructure
J1. Water Treatment Requirements
J2. Loading/Unloading/Storage

Facilities Requirements
J3. Transportation Requirements

K. Instrument and Electrical
K1. Control Philosophy
K2. Logic Diagrams
K3. Electrical Area Classifications
K4. Substation Requirements/Power Sources Identified
K5. Electrical Single Line Diagrams
K6. Instrument and Electrical Specs.

III. EXECUTION APPROACH

L. Procurement Strategy
L1. Identify Long Lead/Critical Equipment and Materials
L2. Procurement Procedures and Plans
L3. Procurement Responsibility Matrix

TABLE 8.1 (*Continued*)

PDRI Sections, Categories, and Elements

M. Deliverables	P. Project Execution Plan
M1. CADD/Model Requirements	P1. Owner Approval Requirements
M2. Deliverables Defined	P2. Engineering/Construction Plan and Approach
M3. Distribution Matrix	
	P3. Shut-Down/Turn-Around Requirements
N. Project Control	P4. Precommissioning Turnover Sequence Requirements
N1. Project Control Requirements	
N2. Project Accounting Requirements	P5. Start-up Requirements
N3. Risk Analysis	P6. Training Requirements

Source: After E. Gibson and P. Dumont, Project Definition Rating Index (PDRI) for industrial projects; *CII Research Report* 113–11. The Construction Industry Institute, 1995.

TABLE 8.2

Comparison of Projects with PDRI—Building Projects Score above and below 200

	PDRI Score	
Performance	**< 200**	**> 200**
Cost	3% below budget	13% above budget
Schedule	3% ahead of schedule	21% behind schedule
Change orders	7% of budget	14% of budget
	($N = 17$)	($N = 61$)

Source: After E. Gibson and M. P. Pappas, *Starting Smart: Key Practices for Developing Scopes of Work for Facility Projects*. Washington, DC: National Academies Press, 2003.

TABLE 8.3

Comparison of Projects with PDRI—Industrial Projects Score above and below 200

	PDRI Score	
Performance	**< 200**	**> 200**
Cost	3% below budget	9% above budget
Schedule	1% ahead of schedule	8% behind schedule
Change orders	6% of budget	8% of budget
	($N = 35$)	($N = 27$)

Source: After E. Gibson and M. P. Pappas, *Starting Smart: Key Practices for Developing Scopes of Work for Facility Projects*. Washington, DC: National Academies Press, 2003.

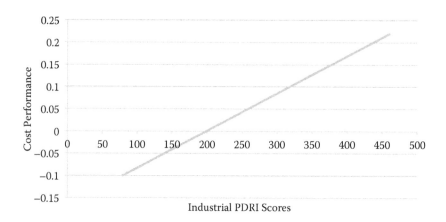

FIGURE 8.1

Cost performance versus industrial PDRI Score. (After Y.-R. Wang and G. E. Gibson, A study of preproject planning and project success using ANN and regression models. In 25th International Symposium on Automation and Robotics in Construction. *ISARC-2008*, pp. 688–696, 2008.)

Wang and Gibson (2008) examined preproject planning data collected from 62 industrial projects and 78 building projects, representing approximately $5 billion in total construction cost. They found that preproject planning has a direct impact on the project success (cost and schedule performance). The graph in Figure 8.1 clearly illustrates the point, showing a linear relationship between the quality of planning and the cost aspect of project success.

There is extensive empirical research in the construction industry linking planning phase completeness and quality to project success in addition to literature confirming the importance of planning from a nonempirical standpoint. This strong body of research along with the strong empirical relationships allow us to conclude

The level of planning completeness is positively correlated with project success in the construction industry.

9

Planning in Information Technology

He who every morning plans the transaction of the day and follows out that plan, carries a thread that will guide him through the maze of the most busy life. But where no plan is laid, where the disposal of time is surrendered merely to the chance of incidence, chaos will soon reign.

Victor Hugo

There have been some notable studies in the software industry where complexity and risk have led to an interest in project management. The reports of high failure rates for software projects and some well-known large failed projects have most likely driven the growth of project management in IT (Sessions, 2009; Standish Group, 2011).

RESEARCH IN IT

Van Genuchten (1991) reviewed the reasons for delays in software development projects based on studies in the late 1980s. He noted that overoptimistic planning, unrealistic project plans, and underestimation of complexity are three of the major reasons for delays of software projects. He reported that in one study 44% of delayed projects cited overoptimistic planning. Another study reported underestimation as a cause of 20% of delayed projects. He also reported that general managers in the affected organizations ranked insufficient front-end planning as the number one reason for schedule slips and cost overruns, and unrealistic project plans was listed as their number two concern. In this study, general managers considered these items to be even more of a factor than did the project managers (PMs). These data strongly point to the benefit of doing more planning and requirements analysis in software development projects. This has become a tenet of software development projects and points to the benefit of more effort in the early stages of projects, including the planning stages.

As Deephouse et al. (1995) note, "A large proportion of systems developed to meet unrealistic commitments may be one cause of the high level of software maintenance expenses reported by many firms" (p. 191). They further state, "When not planned, as in a prototyping strategy, rework can play havoc with schedules and budgets. Substantial rework often introduces new bugs to a system, lowering overall quality" (p. 194). Jorgensen and Boehm (2009) note that challenges still exist in the software project estimation process and less than 10% of projects coming within 10% of initial estimates is the norm. This is after many years of attempting to improve software project estimation (a key component of the planning phase) both through formal methods and better practices within the IT project management community.

Müller and Turner (2001) in reviewing 77 IT projects reported a correlation between postcontract planning (detailed planning after a contract had been signed) and project schedule variance. They report that a quality of postcontract planning that is at least good is required to meet schedule goals. With good planning completed, a zero or positive schedule variance was found whereas negative schedule variances were found otherwise.

Tausworthe (1980) notes the impact of the work breakdown structure (WBS) as an important planning tool with demonstrated benefits on software project success. He defines WBS as "The work breakdown structure (WBS) is an enumeration of all work activities in hierarchic refinements of detail, which organizes work to be done into short, manageable tasks with quantifiable inputs, outputs, schedules, and assigned responsibilities" (p. 181). The WBS is a standard of project management practice and a key deliverable of planning (PMI®, 2008). Lamers (2002) also notes the importance of the WBS to project management from both a theoretical and practical viewpoint. He views it as key to project management practice. Of course, the WBS is one of the deliverables of the planning phase. We can note from this paper that the creation of a WBS is related to the process of defining work packages and therefore project requirements, another key driver of project success (Pinto and Prescott, 1988; Dvir et al., 2003).

I can illustrate these issues from personal experience. When I first started as a project manager at a small dot-com start-up in the late 1990s they were racked by late, out of control projects. They had a stellar customer list from around the world, but when it came to customizing and delivering their product, they were falling down. I had been using formal processes, including formal planning phases, WBSs, Gantt charts, and associated tools for several years by then. I went about my usual process in this new environment. I planned the projects carefully, created detailed WBSs and added appropriate contingency to the developer's numbers. I completed my first project with the company. Low and behold, I came in on time and the project budget was within 3% of estimates. Management was astounded.

CRITICAL SUCCESS FACTORS

Catersels, Helms, and Batenburg (2010) in reviewing critical success factors identified by 129 enterprise resource planning system (ERP) consultants, did not identify project planning in their top 22 critical success factors. However, clear goals and objectives were listed as number four in the list and they would be defined in a planning phase. Interestingly, poor project planning appears as number six in the list of critical failure factors. Weak definitions of requirements and scope was number four in the list of critical failure factors which again is defined in the planning phase. Umble, Haft, and Umble (2003) reported similar findings in ERP implementations: goals not being clearly defined and schedules not being achievable (i.e., poor planning), were two of the ten items noted that would lead to project failure. Tichy and Bascom (2008) reported lack of planning as a cause of project failure in IT as well as other factors that originate in the planning phase such as incomplete requirements and unrealistic expectations. Yeo (2002) in studying critical failure factors in information system projects notes, "[A] significant proportion of problems information system projects faced are related to project planning issues" (p. 246).

Gopal, Mukhopadhyay, and Krishnan (2002) studied application software projects from two Indian software firms, among the largest software firms in India. They found that technical processes, which include requirements management, project planning, product engineering, and software configuration management, showed correlation with total project effort and project time. These items I have identified as being part of the planning phase. They report that these technical processes reduce project effort with a −.714 correlation, rework with −.516 correlation. However, they increase elapsed time with a .702 correlation. This indicates that in these cases good planning may increase elapsed time for the projects but reduces overall effort and rework and would presumably have a positive impact on cost and profitability for the company and therefore overall project success (though that was not what they studied).

Cerpa and Verner (2009), using responses from software practitioners, noted some key software failure factors. The failure factors noted as the most cited, "Delivery date impacted the development process" and "Project underestimated" are items that would have been defined in the planning phase. These two items were identified as problems in 92.9 and 81.4% of failed projects, respectively. As stated in Dvir et al. (2003), in studying defense projects, planning may be a key part of all projects and the impact is only noticed when it is done poorly. This may also be in evidence in IT projects.

EMPIRICAL IT RESEARCH

Deephouse et al. (1995) assessed the effectiveness of software processes on project performance. The results from their paper showed that certain practices, such as project planning, were consistently associated with success, whereas other practices such as process training, stable environment, user contact, design reviews, and prototyping had little impact on the project outcomes. Although the study was designed to focus on process factors and their relationship to success, planning was found to be the leading predictor of meeting targets (efficiency) and quality (which for our purposes is presumed to be related to overall success). The dependency for successful planning was .791 for meeting targets and .228 for quality. It was the top factor in both categories. Although they do note the following caveat about responses: "However, respondents may have had difficulty reporting the effectiveness of planning after the project was complete. They may have allowed the outcome of the project to influence their response as to how well the project was planned. They could have reasoned, 'The project was late, so clearly the plan was not realistic'" (p. 198). They then list as their first key finding, "Effective planning appears to be an important determinant of meeting targets, such as schedules and budgets, and of product quality" (p. 203). The responses of the survey participants at a minimum highlights that they considered planning important.

HOW MUCH TO PLAN

A small number of studies in this area have tried to quantify how much planning should be done for software projects. Posten (1985) stated that in software development projects, testing costs 43% of overall project costs for the projects studied, whereas planning and requirements accounted for only 6% of effort. See Figure 9.1.

He also presented evidence that the earlier defects are identified in the process, the less they cost to fix. See Figure 9.2. The body of literature regarding planning and success in the IT industry supports the importance of planning to success.

Olsen and Swenson (2011) note that "If a fault is detected once the software has been completed, the entire set of analysis, design, development and deployment activities must be performed. If faults are detected earlier in the software development process, fewer steps are required to be repeated. As a result, the costs associated with the fault can be dramatically reduced through early detection" (p. 7). Similarly, Furuyama, Arai, and Lio (1993) conducted a study to measure the effects of stress on software faults. The authors found that 75% of the faults in software development projects were generated during the design phase of the project. Jones (1986) also found that the cost of rework is typically over 50% of very large projects and also that the cost of fixing or reworking software is

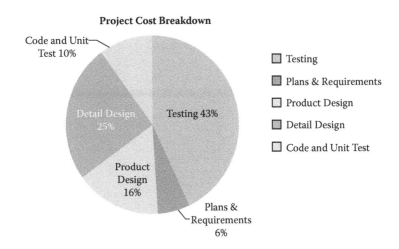

FIGURE 9.1
Project cost breakdown. (After R. M. Posten. *IEEE Software* 2(1), 83–86, 1985.)

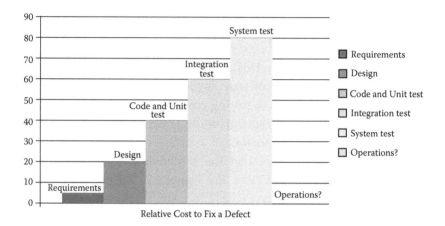

FIGURE 9.2
Relative cost to fix a defect. (After R.M. Posten. *IEEE Software* 2(1), 83–86, 1985.)

much smaller (by factors of 50 to 200) in the earlier phases of the life cycle than in the later phases.

Again, the IT literature points to the importance of planning to success. However, the number of empirical studies in the literature is limited and there is more to learn about the relationship between planning and success. However, we can still conclude

---

The level of planning completeness is positively correlated with project success in the IT industry.

10

Practical Planning Impacts on Success: A Software Development Case Study

To be prepared is half the victory.

Miguel de Cervantes

Below is a description of how to plan and track a small to medium software development project laying out some of the pitfalls. I put this together a few years back but I believe it gives some insight on how the details of quality planning can affect project success. I think you will find some of the advice applies to more than just software projects.

Software development is an area of project management that is notorious for late and unsuccessful projects. I have managed software development projects for a long time and have found some key tips to delivering successful software projects.

Software projects today are being managed differently from the way they were managed in the past. Formal processes are more widely used and the level of experience of the managers involved is better than ever before. Yet many projects are still delivered late, over budget, or both. In the year 2008, 51% of big corporate software development projects surveyed were defined as "challenged" by the Standish Group of research advisers in Dennis, Massachusetts.

Perhaps one reason is that software development poses some unique problems. Two areas where development projects often get into trouble are estimating and tracking.

ESTIMATING FAULTS

Many software development projects fail due to the lack of proper estimates. Software projects are often difficult to estimate for a wide variety of reasons. If estimates are way under, as is the case in many software projects, then there is probably no way to meet the original timelines and budgets.

Why is estimating software projects so hard? Here are some challenge areas:

- Detail: Writing software is about handling a large number of details. Everything must be defined: from the way the main program does its calculations, to writing code, to changing the look of the buttons on the screen, and to writing code that displays the many hundreds of possible errors in a readable manner to the user. When estimating the time to create a piece of code it is very easy to overlook some of this detail.
- Uniqueness: Although attempts have been made to make software code reusable, this effort has had only limited success. Most software projects and most modules within those projects are new to the developers writing them and often new to the world. Most software developers spend their careers writing code to do very new things with every project. As a comparison: on a construction project, if you had only ever put in windows in your career, could you give a good estimate on building a staircase?
- Optimism: Estimates by developers are usually too optimistic. You can trust the estimates of some experienced developers and software architects but I have rarely run into a software person who pads his estimates too much. Often the opposite is the case.
- Unexpected problems: In software, bugs caused by a mistake in single line of code can take days to track down. Other problems can occur when a piece of third-party software doesn't perform as expected and you can spend weeks trying to find out if it is a bug in your code or in the vendor's code. These types of problems can and do occur, are hard to plan for, and can invalidate your project schedule.
- The attitude to planning and process: The attitude that planning and process are a waste of time was prevalent in software development in the past. Some of the comments often heard include the following:
  - Software is creative and we can't interfere with the creative process.
  - Paperwork takes away from development time.

- The schedule is too tight for us to spend time properly designing the solution.
- The documentation is the code. (It isn't.)

Of course, avoiding proper planning only results in late projects or worse: a project that doesn't meet the needs of the end user and ends up being scrapped.

REPLANNING MISHAPS

Even with good estimates and planning, a project can run into problems during execution. Here are some problems to watch for:

- Difficult bugs: Developers will always run into bugs. I have seen a developer spend the better part of two days trying to track down a problem caused by a single character in the wrong spot.
- Gold-plating and tangents: Some staff will start working on a task and quickly use up half the time on the task building the world's greatest configuration utility, a top-of-the-line help system or a module in that sexy new computer language. None of this work may be necessary, planned for, or good for the project. Ensure developers know their deliverables on a weekly basis and don't spend their time doing unnecessary work.
- Drifting away: Not monitoring closely enough can result in your dates slowly drifting to the point where your project will be inevitably late. Nearing the end of the project, it will be too late to take action and get it back on course.

SOLUTIONS

Is there anything a project manager can do to avoid these problems? Here are some tips:

Estimate Right the First Time!

- Ownership has its advantages: Have the developers own the estimates in the project plan. The best way to ensure you have an accurate schedule that the team believes in is to base it on their estimates. Require them to break down and list all their tasks and then estimate the effort. The team lead will also be involved in this process. When the planning is done, get them to agree to the dates. This process will give you better estimates to work with and will motivate the team to meet their deadlines.

- Detailed design: Ensure that each developer has considered design. This doesn't have to be a long process but it needs to be detailed enough to show that the developer has thought through everything that needs to be done. Even a small amount of preplanning and design will improve estimates and improve the quality of the product.

- Design review meetings: Get the team together in a room to discuss the proposed design. What else needs to be done and the risks should also be discussed. You will be surprised at the ideas put forth and the problems avoided.

- Bottom-up is good! The greater the level of detail, the less chance there is that work will be missed in your planning. Work should be broken down to approximately one- to five–daylong tasks. If your project is too large to make that practical, break down the project plan into smaller subproject plans and delegate the estimating. If no one has gone to that level of detail, there is a good chance work will be missed. Put all your detailed tasks together to build the big plan.

- Top-down is bad! Trying to get a plan to fit into dates defined from above without doing detailed estimating can get you into big trouble.

- Previous experience: It is always better to learn from someone else's mistakes than from your own. If you have a similar project that was completed within your organization in the past or if you have access to records from a similar project elsewhere you may use this information as a basis for comparison or for creating your estimates. Best of all is if some of your team have done a similar project in the past. Do what you can to get these people or even consider hiring outside consultants with the right experience. Often, this type of information or experience is not available internally.

- Formal analysis: Formal methods such as function points exist for estimating software projects and are ways of estimating based on an analysis of the requirements for a project. Doing function point analysis requires training but can provide estimates before a detailed design is available and even before the team is assembled. It is not a substitute for detailed design and planning. However, it may be worthwhile for high-risk projects.

Tracking Changes

- Have some contingency: For software projects always ensure you have built in adequate buffers. I would add contingency to individual estimates from junior developers. Developers often don't factor in interruptions, meetings, and phone calls when estimating. In addition, I added an overall contingency to the project as a whole. Even with this added contingency, which may sound like overkill, I still found my project pushing the deadlines.
- Don't rush it: The biggest mistake starting software project managers make is to agree to compress the timelines. This doesn't work in most project management fields and is especially problematic in software development. Rushing through the writing of code is a risky exercise. Small mistakes made in a rush or by a tired developer can be costly. You may find the code written when a developer has stayed an extra six hours late one night ends up taking him or her two days to rewrite during the testing phase. An occasional late night or weekend can help get a project back on track but save it for emergencies; you shouldn't include it in your up-front planning or overdo it.
- See a problem, take action! If dates start to slip, get help for that developer. If you're tracking weekly, you will know early enough to take action. Ensure you know if someone is stuck on a problem and try to help (asking questions can do wonders for expanding developer's thought processes even if you don't have the technical knowledge to give the solution). Get senior team members to help resolve tricky bugs. Get the team lead or a senior developer to help out.
- Track weekly: Meet at least every week to discuss progress and to update your project plan. It will quickly become clear if a developer is having problems or is behind schedule. Group meetings are great for this; developers will not like to report they are late in front of the team and so are motivated to keep on schedule. You should be updating plans and rescheduling as necessary. If your project plan isn't being updated with actuals every week, why not?

- Stop the drift: If you start to see that your estimates and planning are consistently off, you need to take action immediately. Otherwise, the chances of your delivering on time are very low. Renegotiating delivery dates or removing some functionality is much easier to do early in the project than toward the end of the project. Late in the game, capital may have already been spent on the project launch and changing it will be very difficult.

Following these suggestions will improve the consistency of software development projects. Although other things can go wrong and other areas need to be managed, following these tips will maximize your chance of delivering a highly successful software development project.

Does some of the advice above relate to your own projects even if they are not software development? Does this example help to illustrate why the planning phase with its estimating and analysis components is important? It is sometimes useful to understand the challenges and solutions in other fields.

11

Planning and Success in General Project Management

A good plan, violently executed now, is better than a perfect plan next week.

George S. Patton

Now we review the more general project management literature to see what it has to say about planning.

IMPACT OF PLANNING FROM A CONCEPTUAL STANDPOINT

Thomas et al. (2008) state that the front end of the project has received less attention in the project management literature than the subsequent phases that deal with detailed planning and execution: "[T]he most effective team cannot overcome a poor project plan" (p. 105), and projects started down the wrong path in the early stages can lead to the most spectacular project failures. Morris (1998) similarly argued that "The decisions made at the early definition stages set the strategic framework within which the project will subsequently develop. Get it wrong here, and the project will be wrong for a long time" (p. 5). He argued that the role of the initiation phase in defining the project and its influence on project success or project failure poses a strong argument for integrating the initiation phase into the project management domain. This has now largely occurred with the planning and initiation phases now encompassing many project and scope definition tasks. The *PMBOK® Guide* now has the following task: "4.1 Develop Project Charter and 10.1 Identify Stakeholders" (p. 43) as part of the initiating process group of *PMBOK® Guide* (PMI®, 2008).

In my own career, I was brought in to manage an initial outsourcing engagement at a major bank. This engagement was planned to include some radical innovations including the use of open source tools for the majority of systems and the full outsourcing of design and coding to a vendor. The vendor was both a local agency and had offshore resources. Entering the engagement, we were told that there would be some resistance from the regular teams in the department but that because we were being brought in by the CIO, he should be able to overrule any objections. However, what was not understood in the planning of the project was that some of the stakeholder groups did not in fact report to the CIO. On top of that, they were the teams who would need to support the applications after the project was done. In the end, those teams managed to delay, question, and block progress to the point that the project was halted and all the work done to that point scrapped. The stakeholder analysis part of the planning phase was not done effectively and the project was not successful.

Ensuring the project is set up for success prior to the execution stage is important. Jorgensen and Grimstad (2011) note that estimation failures occurring during the planning phase including misestimating required timelines and budgets will of course affect project success. Besner and Hobbs (2011) similarly state, "The most important analysis and initial plans are done during the front-end of the project. If the wrong direction or no clear direction is taken during the early definition phase, it is always difficult to get the project back on track" (p. 21). Munns and Bjeirmi (1996) make a point in stating that a project which is flawed from the early stages is unlikely to be saved by good execution. In fact successful execution may matter only to the project team, and the wider organization will see the project as a failure.

The strategic information systems (SIS) literature also examines the role of planning in enterprise success. Premkumar and King (1991) noted that organizations with higher quality planning processes have better performance in their planning processes which leads to better performance of the IS function and greater contribution of IS to the organization's success. In another paper they note that organizations where IS is considered strategic have better planning processes (Premkumar and King, 1992).

Segars and Grover (1998) and King (1988) confirmed the importance of different aspects of planning to good IS outcomes. Brown (2004) reviewed the SIS planning literature in an attempt to quantify the link between various factors and the quality of the planning. In most of the papers, respondents were asked to rate the quality of the SIS plan produced by the process. However, this research did not link these factors either to project implementation success or other measures of project success. He notes, "The direct effect of the planning process on outcomes … is a glaring omission as many of the benefits of strategic information systems planning (SISP) are intangible and achieved through carrying out the process, rather than producing the tangible plan output" (p. 43). The quality of the plans was routinely being studied but interestingly, their relationship with project success was not.

CRITICAL SUCCESS FACTORS

There is another body of literature studying project success and that is the critical success factor (CSF) literature. The literature studying CSFs does not usually address planning as a CSF although components of the planning phase are noted. Pinto and Slevin (1988) are an exception with project schedule/plans listed as of one of their six key CSFs. Boynton and Zmud (1984) note that planning is a major activity for senior managers. CSFs themselves are a planning tool in their view thus perhaps explaining the fact that planning itself is rarely listed as a CSF. They do conclude that the number one strength of CSFs is that they provide effective support to planning processes. Schultz, Slevin, and Pinto (1987) also reviewed the critical success factor literature prior to 1987. They found that four out of five of the key studies to that point cited planning tasks as CSFs. They also note that four of the nine most critical CSFs they identify are also part of the planning phase. It is interesting to note that subsequent CSF research has focused less on planning. Perhaps it is because planning has now become more a normalized part of projects and included as a default in all projects.

More recent studies of CSFs have focused less on planning as a CSF. But interestingly poor planning is often noted as a critical failure factor (Umble et al., 2003; Catersels et al., 2010; Yeo, 2002). Ewusi-Mensah (1997) noted that key factors in canceled projects are poor project goals, poor technology infrastructure, and escalating costs and timelines. These would normally be analyzed and addressed in a thorough planning phase. Yeo (2002) states that of the issues of influence noted in this study of critical failure factors, project planning was ranked as number one. This research again supports the contention that planning is hygiene to project success (Turner and Müller, 2003). From Yeo's and other research, it could be one of the more critical hygiene factors.

Many CSFs mentioned in this body of literature contain deliverables, which are created in the planning phase, often near the top of the lists. White and Fortune (2002) note in their survey of 236 project managers that clear goals and objectives and realistic schedule were the two most mentioned critical success factors out of the 22 identified. Again these are factors that are defined in the planning phase. One can argue that goals and objects are defined at the project inception but they would also be expected to be clarified and detailed during the planning, before execution. Pankratz and Loebbecke (2011) noted that all 11 participants in their interviews mentioned planning, monitoring, and control as a project success factor, one of only two items to be mentioned by all participants (the other was team members' motivation).

Blomquist and colleagues (2010) state, "Plans are a cornerstone of any project; consequently, planning is a dominant activity within a project context" (p. 11). This is, of course, a recurring theme: projects and project management are about planning and controlling to ensure successful project deliverables. Planning is inherently important to project success or one could argue project management would not exist.

In the 1990s I started a company with a partner. My plan was to pool our investments and build a multiplayer game. I had a full design for the game with a world and gameplay all laid out. At that time, multiplayer web games were in their infancy. The pricing model had not evolved for Internet games so I wasn't sure how we would make money and I was not certain if it was even technically feasible. We soon scrapped that plan and chose instead to work on getting venture financing. This eventually became our focus and we spent our investment dollars on building demos and prototypes in hopes of getting that six- or seven-figure financing agreement. But it never arrived and we ran through our money. A few years later, others had managed to sell their embryonic games for millions of dollars. My assessment now is that the original plan was the right one. By changing our original plan we may have missed a great opportunity.

PLANNING TOOLS' EFFECT ON PROJECT SUCCESS

Besner and Hobbs (2006) in studying project management tools and success found that five of the eight "super tools" most clearly linked to project success are used or created during the planning phase:

- Software for task scheduling
- Scope statement
- Requirements analysis
- Gantt chart
- Kickoff meeting

The remaining "super tools" are as follows:

- Lessons learned/postmortem
- Progress report
- Change requests

This is significant; although planning typically is a fraction of overall project effort, it has a disproportionate impact on project success.

In a paper published subsequently, Besner and Hobbs (2011) found that initial planning was the number one used tool set reported by the 744 respondents. In their definition, initial planning consisted of the following:

- Kickoff meeting
- Milestone planning
- Scope statement
- Work breakdown structure
- Project charter
- Responsibility assignment matrix
- Communication plan

PLANNING PHASE COMPONENT COMPLETENESS/ QUALITY AND PROJECT SUCCESS

Pinto and Prescott (1988) analyzed responses from 408 project managers and team members and examined the impact of critical success factors on project success. They found that a schedule or plan had a correlation of 0.47 with project success and detailed technical tasks had a correlation of 0.57 and mission definition a correlation of 0.70. These success factors are products of what I have defined as the planning phase. This study did not use regression to examine further the impact of these factors on project success, only the correlation. However, schedule/plan did have an R^2 relationship of .57 with a successful execution phase, mission had $R^2 = .50$, and technical tasks had an $R^2 = .59$: all very significant values. Unfortunately, overall project success was not analyzed, only the impact on execution.

Shenhar et al. (2002), using multivariate analysis, state that design considerations have a major impact on the success of high-uncertainty projects and in high-technology/high-uncertainty projects, this should be a major management focus. Another aspect of the planning phase, the WBS (work breakdown structure) creation, is also critical in high-uncertainty projects although it is less critical in low-uncertainty projects. The creation of a detailed WBS may itself be a key part of planning as it can be important in helping the team ensure all project tasks have been considered. The WBS is also noted as the backbone of proper planning, execution, and control of a project in the literature (Bachy and Hameri, 1997; Zwikael and Globerson, 2006). McFarlan (1981) notes, "Project life cycle planning concepts, with their focus on defining tasks and budgeting resources against them, force the team to develop a thorough and detailed plan (exposing areas of soft thinking in the process)" (p. 147) and this is a key aspect and benefit of building a complete WBS.

Shenhar et al. (2002) found that certain aspects of the planning and control phases (WBS creation) as well as what they termed policy and design considerations (design cycles) are linked to success in all types of projects. Almost all the factors noted were linked to success in high-uncertainty projects. From their research, it also was clear how documentation produced in the planning phase is critical to project success.

I once managed a program in a medium-sized company with no tradition of project management. I was given a team consisting of specialists with good subject knowledge but without formal project management experience. My role was to build a program structure. This was very necessary: the program had a total team size in the hundreds. I set about training those project managers who weren't familiar with it, how to use MS Project. I then mandated all of them to create a WBS by going to each developer and getting component tasks and the time required to complete all of them. With a little bit of training and discipline, we were able to build individual project plans and combine them into a consolidated program plan. It wasn't easy but it worked. The program was delivered on time and on budget and considered highly successful by the end client.

Zwikael (2009), through analysis of 783 questionnaires, studied the contribution of the *PMBOK® Guide*'s nine knowledge areas to project success. He reported that the knowledge areas related to the planning phase had the highest impact on project success "the more frequently planning processes—which are related to these Knowledge Areas—are performed, the better project success is." Conversely, "Cost and Procurement are the Knowledge Areas that contribute least to project success, maybe because they are practiced mainly during project execution" (p. 98). He found that the top three knowledge areas found to be most correlated with project success are typically associated with planning phases. As well, the knowledge areas most linked to project success are also the ones most commonly performed in the planning phase. Table 11.1 shows the knowledge areas found to be most correlated with project success and the top three are typically associated with planning phases.

From this table, it is clear that the knowledge areas most linked to project success are also the ones most commonly performed in the planning phase. Of the top five knowledge areas ranked by contribution to project success (time, risk, scope, human resources, and integration) all also appear in the top six of knowledge areas used during the planning phase.

TABLE 11.1

Importance of the Nine Knowledge Areas to Project Success

Knowledge Area	Contribution to Project Success (Ranked)	Use during the Planning Phase
Time	1	High
Risk	2	Medium
Scope	3	High
Human resources	4	Medium
Integration	5	High
Quality	6	Low
Communications	7	Low
Cost	8	Medium
Procurement	9	Low

Source: After O. Zwikael, *Project Management Journal* 40: 94–103, 2009.

Zwikael concludes by advising that project managers should invest more effort in identifying project activities, developing Gantt charts, and identifying the critical path or the critical chain of a project, and using these in cooperation with key stakeholders to improve project success.

Koskela and Howell (2002) made similar observations about the *PMBOK® Guide*. They note, "The planning processes dominate the scene in the *PMBOK® Guide*: in addition to the ten planning processes, there is only one executing process and two controlling processes. The emphasis is on planning, with little offered on executing especially" (p. 4). Although the *PMBOK® Guide* has changed in the past 10 years, this is still largely true.

Pinto and Prescott (1990) in a seminal study of 408 project managers found that when internal measures of project efficiency are compared to critical success factors, planning is perceived to be of high importance at the initial stages but is overtaken by tactical issues. However, for external success measures or overall project success, planning factors dominate throughout the project life cycle. Planning was found to have the greatest impact on the following success factors: perceived value of the project ($R^2 = .35$) and client satisfaction ($R^2 = .39$). This result highlights the impact of planning on success in projects.

Johnson, Boucher, and Connors (2001) note the 10 factors most associated with project success (as captured in the Chaos report by the Standish group; Table 11.2); four of ten are related to the planning phase: clear

TABLE 11.2

Recipe for Success: CHAOS 10

Confidence Level	Success Factors (Weighting)
Executive support	18
User involvement	16
Experienced project manager	14
Clear business objectives	12
Minimized scope	10
Standard software infrastructure	8
Firm basic requirements	6
Formal methodology	6
Reliable estimates	5
Other criteria	5

Source: After J. Johnson, K. Boucher, and K. Connors, *Software Magazine* 7: 1–9, 2001.

business objectives, minimized scope, firm basic requirements, and reliable estimates. Two others, which are defined during planning, can be partially related to the planning phase: formal methodology and user involvement (p. 3).

They also note that project success rates as measured by the Standish group increased between 1994 and 2000 with more projects succeeding and fewer projects failing. One could speculate this could be attributed to an improvement in project management practices during that period.

PLANNING PHASE COMPLETENESS AND PROJECT SUCCESS

Shenhar (2001) in a survey of 127 project managers notes better planning is the norm in high and super-high technology projects. This was found to apply consistently to the deliverables normally produced in the planning phase. Managers appear to believe that more complex and higher risk projects require a greater planning effort and more complete planning deliverables.

Dvir and Lechler (2004) in reviewing data from 448 projects collected in Germany found a correlation between the quality of planning, project efficiency, and customer satisfaction. Quality of planning had a +.35 impact on R^2 for efficiency and a +.39 impact on R^2 for customer satisfaction. Yetton et al. (2000) from a survey with 71 participants found a strong correlation between planning and budgets. They report a correlation coefficient of –0.39 between planning and budget variances. In other words, the better the planning, the lower is the likelihood of budget overruns. They also found that planning had an indirect positive impact by reducing team instability as well as the direct impact on budgets. They noted, "Good planning facilitates and assists a stable and skilled project team to perform more efficiently" (p. 284).

Salomo, Weise, and Gemünden (2007) studied the relationship between planning and new product development projects. They found that project risk management and project planning together had an R^2 impact of .28, although the contribution of project planning was not significant. I consider risk planning part of the planning phase in this review, therefore the overall R^2 = .28 for the planning phase to success. In addition, they reported process formality and goal clarity gave an R^2 = .33 to success which are two items defined in the planning phase.

PLANNING PHASE EFFORT AND PROJECT SUCCESS

In a well-known paper on the topic, Dvir, Raz, and Shenhar (2003) studied 110 defense-related projects over 20 years. They noted the correlation between aspects of the planning phase and project success. Although the paper spoke of planning phase efforts, what was being measured was the quality of three types of planning deliverables: functional requirements, technical specifications, and planning processes and procedures. They did not find a strong correlation with planning procedures, however. Therefore the project planning and WBS effort was found to be less important to project success than defining functional and technical requirements of the project. Both classes of activities are part of what I have defined as the planning phase. They reported, however, that two aspects of project planning such as defining functional requirements and time spent on technical specifications are correlated with perceived project success. The correlation was $R^2 = .297$ for functional requirements and .256 for technical requirements.

They also state, "[A]lthough planning does not guarantee project success, lack of planning will probably guarantee failure" (p. 94). This statement is broadly supported by the CSF literature discussed previously (Umble et al., 2003; Catersels et al., 2010; Yeo, 2002). Based on the body of literature, we can therefore generalize the previous conclusion to include projects outside the construction and IT industries:

The quality of planning is positively correlated with perceived project success in projects in general.

The main conclusions Dvir et al. (2003) reach is that "There is a significant positive relationship between the amount of effort invested in defining the goals of the project and the functional requirements and technical specifications of the product on one hand, and project success on the other" (pp. 94–95). They conclude by stating that no reasonable effort should be spared in the early stages of a project to define the project goals and requirements properly.

Zwikael and Globerson (2006), in data collected from 280 project managers, noted the following: "Once processes are performed correctly at the planning phase, it will be easier for the project manager to continue

[to] manage the other project phases at the same level of quality, until the project's successful conclusion" (p. 698). They conclude that there is a relationship between project planning effort and project success.

This research is consistent with the previously detailed studies where Posten (1985) reported on the planning effort required for successful software projects and Choma and Bhat (2010) found planning effort and success in a variety of industrial projects. Projects with too little planning were less successful. More on their study later.

Therefore it is clear from the preponderance of the literature that we can report

---

Planning quality is associated with both project efficiency and overall project success.

CONCLUSION

What appears to be clear is that activities that I defined as part of the planning phase such as requirements definition, scope definition, and technical analyses are important to project success. One can argue that the planning phase is not as important as what comes with it. As Eisenhower is said to have stated, "In preparing for battle I have always found that plans are useless, but planning is indispensable." Activities such as the creation of the detailed WBS, stakeholder analysis, requirements definition, and risk analysis all take place in the planning phase and have all been found critical to project success.

It is clear that the activities that occur prior to execution and along with planning are important to project success although the narrowly defined planning processes of creating timelines and Gantt charts are not as critical to project success as are some other activities (Dvir et al., 2003). Catersels et al. (2010) and Poon et al. (2011) found that high-level planning was the least important of the CSFs that they studied. Turner and Müller (2003) also note that "There is growing evidence that competence in the traditional areas of the project management body of knowledge are essential entry tickets to the game of project management, but they do not lead to superior performance. They are hygiene factors, necessary conditions for project management performance, but they are not competitive factors for which improved competence leads to superior project performance" (p. 6). Perhaps it is the detailed planning and analysis that occurs during the planning phase that ensures the team has covered all aspects of the project. This may have the most impact on project success.

12

Consensus on Planning and Success

Plans are only good intentions unless they immediately degenerate into hard work.

Peter Drucker

LITERATURE SUMMARY

I recall on one consulting assignment, my program manager asked me how much I could take on. Being young and cocky, I told her that she could keep assigning me projects and I would let her know when I was overloaded. Soon, I was managing eight projects. I was doing the right things in general: holding team meetings, scheduling meetings with stakeholders, starting detailed analyses, and providing status reports. But it soon became difficult to keep some of the projects straight in my mind. I would mix up the names and acronyms as I moved from meeting to meeting. It started to appear to the project stakeholders that I was not on top of things even though I believed everything was happening as it needed to happen. Maybe they were even right. I perhaps was not managing the analysis and planning the way that I should have. The main aspects of the projects were on track but the perception with some stakeholders was I was not fully committed. And perception is often reality. My contract with that client was cut.

There have been numerous studies that link project success to the planning phase either through analysis of project results or as a statement of observations. This list is summarized in column 1 of Table 12.1. In general, the research is consistent: the majority of studies, with a few outliers, state planning is important to project success.

TABLE 12.1

Summary of Positions of Reviewed Literature on Project Planning

Positive Empirical Relationship between Planning and Success	Conceptual Positive Relationship between Planning and Success	No Relationship between Planning and Success	Conceptual Negative Relationship between Planning and Success	Empirical Negative Relationship between Planning and Success
Pinto and Prescott (1988)	Tausworthe (1980)	Flyvbjerg et al. (2002)	Bart (1993)	Choma and Bhat (2010)
Pinto and Prescott (1990)	Chatzoglou and Macaulay (1996)		Anderson (1996)	
Hamilton and Gibson (1996)	Munns and Bjeirmi (1996)		Boehm (1996)	
Deephouse et al. (1996)	Ewusi-Mensah (1997)		Zwikael and Globerson (2006)	
Müller and Turner (2001)	Morris (1998)			
Shenhar et al. (2002)	Johnson et al. (2001)		Aubrey et al. (2008)	
Dvir et al. (2003)	Shenhar (2001)		Collyer et al. (2010)	
Gibson and Pappas (2003)	Yeo (2002)		Poon et al. (2011)	
Dvir and Lechler (2004)	Umble et al. (2003)			
Gibson et al. (2006)	Ceschi (2005)			
Zwikael and Globerson (2006)	Mann and Maurer (2005)			
Besner and Hobbs (2006)	Besner and Hobbs (2006)			
Salomo et al. (2007)	Smits (2006)			
Wang and Gibson (2008)	Thomas et al. (2008)			
Zwikael (2009)	Shehu and Akintoye (2009)			
Choma and Bhat (2010)	Zwikael (2009)			
	Blomquist et al. (2010)			
	Collyer et al. (2010)			
	Catersels et al. (2010),			
	Besner and Hobbs (2011)			
	Pankratz and Loebbecke (2011)			

Source: After P. Serrador, *Journal of Modern Project Management* 2: 28–39, 2013.

From this table, we can see that the preponderance of the literature has found that planning and the level of completeness of planning are important for project success. From the literature review alone we can answer the question: Is planning important for project success? The answer is yes.

EMPIRICAL RESULTS

Now we can look at how the quality of planning can affect project success. As we discussed, the coefficient of determination R^2 provides a measure of how well future outcomes are likely to be predicted by a variable (in this case, measures of planning). We can compare the calculated R^2 values from a number of papers.

Some of the papers reviewed in the literature review reported empirical data. This allows us to embark on a high-level meta-analysis of those results. Meta-analysis is analyzing and contrasting results from different studies, to try to identify patterns in the study results. The technique employed was to examine reported R and R^2 results, categorize them as relating primarily to efficiency or to overall success, and review the means. Conducting a rigorous mathematical analysis was not possible given the varied nature of the source documents: different industries, different methodologies, and cross-functional projects in some studies and not others. A high-level meta-analysis reviewing the means was completed instead, as shown in Table 12.2. These studies used different methodologies and even different definitions of planning and success. However, the results appear to be surprisingly consistent. We can report

---

As an approximation, research shows an average value of $R^2 = .33$ correlation with efficiency and $R^2 = .34$ with success.

This indicates a significant impact on project success. Although 33% may not seem like much, it will have a substantial impact. Figure 12.1 graphically illustrates how big an impact $R^2 = .33$ can have. This can turn a successful project into an unsuccessful project or turn a mediocre project into a highly successful one.

If we compare this .33 effect to the approximately 20–33% effort spent on planning reported by Wideman (2000), Nobelius and Trygg (2002),

TABLE 12.2

Meta-Analysis Summary of Empirical Results

Study	Empirical Relationship	Aggregate	Normalized to R^2	
			Efficiency	Overall Success
Pinto and Prescott (1990)	Planning found to have the greatest impact on success factors. Perceived value of the project ($R^2 = .35$). Client satisfaction ($R^2 = .39$).	$R^2 = .35$ $R^2 = .39$ Average $R^2 = .37$	$R^2 = .37$	$R^2 = .39$
Hamilton and Gibson (1996)	The top third best planned projects had an 82% chance of meeting financial goals whereas only 66% of projects in the lower third did. Similar results were seen in these projects' results relating to schedule performance and design goals met.			
Yetton et al. (2000)	They found a correlation coefficient of −0.39 between planning and budget variances.			
Deephouse et al. (1995)	The dependency for successful planning was .791 for meeting targets and .228 for quality.	$R^2 = .625$ $R^2 = .052$ Average $R^2 = .34$	$R^2 = .34$	
Dvir et al. (2003)	Meeting the planning goals is correlated .570 to overall project success measures.	$R^2 = .32$		$R^2 = .32$
Dvir and Lechler (2004)	Quality of planning had a +.35 impact on R^2 for efficiency and a +.39 impact on R^2 for customer satisfaction.	$R^2 = .35$ $R^2 = .39$ Average $R^2 = .37$	$R^2 = .35$	$R^2 = .39$

Study				
Zwikael and Globerson (2006)	Planning quality correlates as follows: R = .52 for cost. R = .53 schedule. R = .57 technical performance. R = .51 customer satisfaction.	$R^2 = .27$ $R^2 = .28$ $R^2 = .32$ $R^2 = .26$ Average $R^2 = .28$	$R^2 = .28$	$R^2 = .29$
Gibson et al. (2006)	$R^2 = .42$ correlation between planning completeness and project success.	$R^2 = .42$	$R^2 = .42$	
Salomo et al. (2007)	$R^2 = .27$ between project planning/risk planning and innovation success. $R^2 = .33$ between goal clarity/process formality and innovation success.	$R^2 = .27$ $R^2 = .33$ Average $R^2 = .30$		$R^2 = .30$
Wang and Gibson (2008)	PDRI score of a building construction project is related to project cost and schedule success (R = .475).	$R^2 = .23$	$R^2 = .23$	
Overall Average		$R^2 = .33$	$R^2 = .33$	$R^2 = .34$

Source: After P. Serrador, *Journal of Modern Project Management* 2: 28–39, 2013.

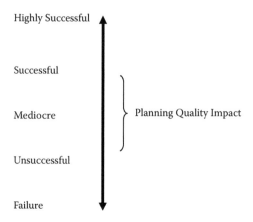

FIGURE 12.1
Impact of $R^2 = .33$.

and Chatzoglou and Macaulay (1996), there appears to be a clear return on this investment in terms of project success.

However, whether there is an ideal amount of effort that should be spent planning in a project is still an area for further discussion. And I cover that area in Chapters 16 and 17.

13

Planning Deliverable Quality and Success

Action is the foundational key to all success.

Pablo Picasso

But how does planning quality relate to success based on these new data? The literature that I reviewed in Chapters 11 and 12 showed the impact of the quality of the planning phase and planning deliverables on success. Did my research confirm those findings? I now review my results from a quality of planning perspective.

ORIGINAL RESEARCH

I captured a number of project characteristics in my survey which I studied as moderating variables to my main relationship, as shown in Figure 13.1. The project variables that I collected are shown in Table 13.1.

The original intent of gathering these variables was to examine them as moderators and study their impact individually. However, after performing a factor analysis to see if some of the moderators were related, it became clear that four of them were connected and described an underlying factor. The details of the full factor analysis can be found in Appendix B.

Factor 2 = mean of the following four responses:
1. Detail level of the WBS
2. Quality of goals/vision
3. Degree of stakeholder engagement
4. Experience level of team

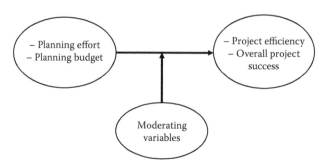

FIGURE 13.1
Main research constructs.

When I looked at this factor and tried to understand its implications for project success, it seemed clear to me that this factor was really related to the quality of planning on a project. I therefore named it the planning factor and considered it a measure of the quality of planning on the project. Some components can be clearly seen to be the result of a thorough analysis and planning exercise: detail level of the WBS and quality of goals/vision. Another component can be seen as an important input to a good planning effort: degree of stakeholder engagement. The experience level of the team does not immediately come to mind as a planning related variable. However, based on the factor analysis, it is related to planning quality. One can speculate that this may be because a better planning cycle may allow the selection of a more effective team or that more experienced teams complete more effective planning.

QUALITY OF PLANNING AND SUCCESS

Now that the meaning of the factor had been defined, I next completed a regression of success versus the planning factor, as shown in Table 13.2. The table shows a statistically significant relationship with a very low $p < .0001$ between the planning factor and overall success. In addition, there is a strong R^2 of .265. This result is in broad agreement with the R^2 reported in the literature of .34. The planning factor calculated here is a measure of planning quality similar to what was studied in previous research. However, it is not as complete, therefore a lower R^2 is to be expected.

This analysis is quite theoretical but does also relate to the real world. I recall an occasion when an outsourced project was in trouble. It was fixed

TABLE 13.1

Moderating Variables in the Planning/Success Relationship

Moderating Variables	
	1. Project team size How large was the project team (full-time staff equivalent)?
	2. Project complexity Rate the complexity of the project.
	3. Project length What was the project length (full life cycle)?
	4. WBS (work breakdown structure) used Rate the detail level of the WBS in the project.
	5. Goals/vision quality Rate the applicability/quality of the vision statement or project goal definition for the project.
	6. Novelty to the organization How new is this type of project to the organization?
	7. Internal versus vendor based What percentage of the project was completed by vendors?
	8. Industry In what industry was the project? (Choose the best fit.)
	9. Geographic location of project What was the geographic location of the project?
	10. Local versus remote team Where were the team members located? Choose the option that best fits the majority of team members.
	11. Level of use of technology Low tech indicates none or very mature technology where super-high tech indicates the use or development of completely new technology.
	12. New product versus maintenance Does this project involve developing a new product, installation, or system or is it related to maintenance of what already exists?
	13. Experience level of team How experienced was the project team?
	14. Degree of stakeholder engagement How engaged were the key stakeholders for the project?
	15. Methodology type How much of the project was done using agile or iterative techniques? (100 = fully agile, 0 = fully waterfall, 50 = an equal mix of agile and waterfall techniques.)

TABLE 13.2

Regression Analysis for Planning Factor versus Success Measure

Regression Summary for Planning Factor versus Success Measure				
	R	$R^2 =$	Number of Projects	p-Level
Planning Factor	−0.515	.265	1,386	0.000

price but the offshore company was having to throw more and more people at the project. Fixed price means the offshore partner agrees to complete the work for a fixed price regardless of their cost. This shifts some of the risk for the project to the offshore partner. However, they will add contingency on to their quote and may take a very cautious approach. As well, in some cases, fixed price may not really be fixed price.

Time and materials means the client only pays for the work they do but the client takes on some of the risk. If the project gets out of control and the amount of work greatly increases, so will the costs. Also, with distant staff, how can the PM be certain about the hours they are working?

In this particular project, the vendor added more people to the team but still they were going nowhere. The offshore team had designed the approach and they had underestimated the complexity of the task. So what could I do? The offshore team threatened to walk away to cut their losses. I had promised these changes to the business partners and canceling the project wouldn't reflect well either. We ended up renegotiating the contract for a new price with reduced scope. Thus, fixed price is not always fixed price. In the end both the client and the vendor had failed to do their planning and analysis properly.

PLANNING AND EFFICIENCY

Now, let's continue with the regression analysis. If we look specifically at the planning factor's impact on project efficiency measures, the relationship is still there but has different characteristics, as shown in Table 13.3. The table shows a statistically significant relationship with a very low $p < .0001$. However, in this case, R^2 is .14 which is less than half the effect reported in the literature. This is an interesting result. One would expect that quality of planning would allow projects to deliver more successfully

TABLE 13.3

Regression Analysis for Planning Factor versus Efficiency Measure

	R	$R^2 =$	Number of Projects	p-Level
Regression Summary for Planning Factor versus Efficiency Measure				
Planning Factor	.376	.141	1,386	0.000

TABLE 13.4

Regression Analysis for Planning Factor versus Success Measure

	R	$R^2 =$	Number of Projects	p-Level
Regression Summary for Planning Factor versus Stakeholder Success Measure				
Planning Factor	.525	.275	1,386	0.000

on time, on scope, and on budget. But the quality of planning seems to have a greater impact on the broader success measures. We can speculate that perhaps more time spent increasing planning quality improves the final outcomes of the project but has a somewhat negative impact on meeting time and budget goals.

Then I examined the nonefficiency success measures; the results are again different. Table 13.4 shows a statistically significant relationship with a very low $p < .000$. However, in this case, R^2 is .28.

This research has also confirmed the findings in the literature that the quality of the planning outputs is very important to project success (Pinto and Prescott, 1990; Dvir et al., 2003; Gibson et al., 2006). The relationship between planning quality and project success with an R^2 in the .34 range according to the literature and $R^2 = .28$ as found in this research shows that the quality of planning is responsible for up to a third of a project's success. The quality of the planning outputs is obviously a key factor for project success that project managers need to be aware of when designing their planning phase.

Therefore we can confirm

Quality of planning is clearly an important factor in project success and has a clear predictive relationship with success.

14

How Much to Plan

Give me six hours to chop down a tree and I will spend the first four sharpening the ax.

Abraham Lincoln

So it is clear that planning is important to success. Now we should ask how much planning should be done in projects. Surprisingly little research has been done in this area, although one would assume that this kind of guidance would be useful for project managers and researchers. Let's review the research that is available in this area.

HOW MUCH TO PLAN IN IT

Daly (1977) in a paper on software development stated that for software projects, planning (i.e., schedule planning) should be 2% of total project cost and specifications should cover 10% of the total cost. Final design was to take a further 40% of cost. However, the practice and technology has changed in the more than 30 years since this paper was written. Design does not need to be all done up front anymore as interactive compilers now allow low-level design and coding to be completed together. The 2% of effort on planning appears quite low. Let's look at more research.

Chatzoglou and Macaulay (1996) in reviewing data from IT projects in the United Kingdom asked

> How much planning is enough? There is no standard one single answer to this question. The right amount of planning depends on the size of the project, the size of the development team and the purpose of the plan. Plans are

made at a number of points in a development project for varying reasons. The timing of planning tasks depends on the organisational environment and the nature of the particular project. (p. 175)

They also outline a rule of thumb for planning effort: The three-times-programming rule and the life cycle stage model. "One rule of thumb that has been in use for a long time is the 3-times-programming rule. With this method, one estimates how long it would take to program the system and then multiply by three to estimate how much effort it will take to deliver a tested, documented system" (p. 183). Although Chatzoglou and Macaulay do not specify this, traditionally software development testing is estimated to take roughly an equal amount of effort as development (Kaner, Falk, and Nguyen, 1999). This leaves one third of total effort for the planning phase and the other miscellaneous tasks. These ratios are not only noted in the academic literature but they are known and practiced in industry.

An experienced and successful project manager friend of mine introduced me to these ratios when I was working for a leading North American insurance company. I was uncertain about the estimates I had put together due to the relative inexperience of the team. I reviewed the schedules with these ratios in mind. After additional vetting and redoing some of the estimates, I had a final plan that came closer to the expected ratios. My friend's advice was right; the project came in within 5% of time and budget.

Ten years earlier than Chatzoglou and Macaulay, Posten (1985), had stated that plans and requirements should be 6% of the project cost, product design should be 16%, and detailed design 25%. This unfortunately does not give clear guidance on how long the planning phase itself should be.

It is interesting to note that empirical guidance on how much time to spend on each phase was more common in the earlier years of software development project management. Studies on this topic subsequently have not specified how much time to spend on each phase. It's not clear whether this is because this guidance was found not to be effective, perhaps the diversity of technology projects increased, or interest was lost in this area of research.

OTHER INDUSTRIES

Nobelius and Trygg (2002) analyzed front-end activities (similar to planning phase) for three projects in varied industries. They reported that in these three case studies, front-end activities made up at least 20% of the project time. Similarly, Wideman (2000), using data derived from a fairly large sample of building projects with budgets up to several million dollars, states that the typical effort spent in the planning phase in construction projects is approximately 20% of the total work hours. He states that work hours typically make up 40% of total costs. Therefore for building projects approximately 20% of person-hours and 8% of budget is spent planning.

Is the amount of time and effort spent planning important to success? Choma and Bhat (2010) analyzed 73 construction and industrial projects from 49 different organizations with total costs between $10 million and $500 million. They report, "As would be expected, the projects with the worst results were those that were missing important planning components at authorization, such as a detailed schedule, complete basic engineering approved by operations and maintenance, a risk analysis with mitigation plans, complete responsibility matrix, etc." (p. 5). However, they did not find a correlation between time spent in the planning phase front-end loading (FEL) and project success. In fact, "[T]he projects in this sample that took longer in planning had the worst results. On average, the Worst Projects had an FEL phase that was roughly 71 percent longer than the average for the Best Projects. Thus, the time spent in FEL does not determine the quality of planning; rather, it is the deliverables completed in FEL that are most correlated with results" (p. 7). Their analysis points to either that too much planning can be negative to project success or that a planning phase that lasts too long can be an indicator of a problem project. This could also be related to the analysis paralysis effect when so much analysis takes place that no actual work is started, or it is started later than optimal (Rosenberg and Scott, 1999).

For one late program I was involved with, it was clear almost no formal project planning had been completed when I joined. This was probably due to the fact that the majority of the project managers were business-people with no project management training. The new program manager brought in a template from a previous successful program. However, the template was very, very detailed. Weeks were spent by the project managers and team understanding and completing the template line by line. Although there were benefits to ensuring no tasks were missed in the project, this was very likely a case of too much planning and at the wrong time. Instead of getting ready for execution, hours were spent poring over the details that wouldn't be important for more than a year. Other work fell by the wayside. In that case, this was not a good use of time at a critical point in the project.

CONCLUSION

There has been relatively little research on what the optimum percentage of efforts is for the planning phase and the literature that exists is not recent and may not be fully consistent. The literature does point to a recommended planning effort of 20%–33% but these recommendations are not based on solid empirical research. Therefore, it is difficult to make definitive conclusions in this area and further research is warranted.

In the next chapter, I detail my own research in this area.

15

Planning Phase Effort and Success

Success is dependent on effort.

Sophocles

Now I can start to answer the question that was the original reason for this research: How does the time spent planning affect project success? The literature on this topic only peripherally touch on the answer to this question. From the data I collected and the questions I asked, I hoped to change that.

PLANNING EFFORT VERSUS SUCCESS

To start the analysis, I examined the relationship between planning effort index and project success rating, Table 15.1. I removed data elements that had invalid planning amounts prior to starting my data analysis which is detailed in Appendix A. We can see that in general the planning index increases within the success category. The exception is the failure category that shows the highest mean planning effort index of any group. The ANOVA analysis does not show a statistically significant relationship. By looking at these means it appears that a simple linear relationship is unlikely to exist.

These data were now plotted to get a visual picture of the relationship, Figure 15.1. Looking at this graph, we can see the lowest amount of effort was typically spent on projects deemed not fully successful. In this case, one can hypothesize that inadequate planning affected project success. Projects deemed outright failures reported the mean highest percentages of upfront project planning. This is an interesting finding and in keeping

TABLE 15.1

Planning Effort Index and Project Success Rating
for All Projects

	Planning Effort Index	Number of Projects
Failure	0.166	98
Not Fully Successful	0.143	259
Mixed	0.152	345
Successful	0.154	451
Very Successful	0.158	233
All Groups	0.153	1,386
p(F)	*0.178*	

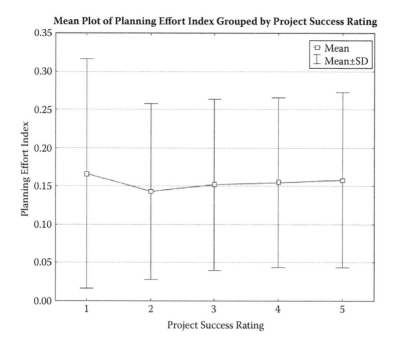

FIGURE 15.1
Mean plot of planning effort index by project success rating with error bars.

with the effect reported by Choma and Bhat (2010). Projects in trouble may spend extra effort in the planning phase due to a very complex or challenging project or problems early in the project.

These types of projects may also be canceled before full completion which would inflate the planning indexes. Some of the comments provided by my respondents appear to indicate this: "Approximately 60% through engineering design the cost and schedule were deemed too great to continue and the project was suspended," "The second (unsuccessful) project I listed was cancelled by client since they merged with another company," and "Failed project was primarily caused by inability of key stakeholders to agree on business rules." These comments point to projects spending more time than usual in the planning phase.

NONLINEAR RELATIONSHIPS

Based on Figure 15.1, it was decided to review the data with an assumption that the relationship between the effort index and project success is not linear but could be polynomial in nature: that is, a curve rather than a straight line.

Figure 15.2 was produced via a scatterplot analysis by a statistics program. There is clearly a quadratic relationship between the planning effort index and the success measure. This fits with the position that if

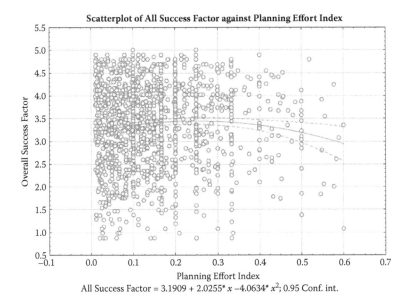

All Success Factor = $3.1909 + 2.0255^* x - 4.0634^* x^2$; 0.95 Conf. int.

FIGURE 15.2
Scatterplot and curve fitting for success measure (overall success factor) versus planning effort index.

TABLE 15.2

Nonlinear Regression Analysis of Planning Effort Index
versus Success Measure

	B_x	R^2	p-Level
Nonlinear Regression Summary for Planning Effort Index versus Success Measure			
Intercept	3.191		.000
Planning effort index	2.026		.001
Planning effort index^2	−4.063		.003
Overall model		*.006*	*.0059*

a project spends too much effort in the planning phase, too much of the overall budget will be spent and the project will start later than it would otherwise (Chatzoglou and Macaulay, 1996). The project would end up being less successful overall. Conversely, a project that spends too little up-front time planning will also be less successful (Dvir et al., 2003). Therefore an inverse U-curve fits with this proposition and the findings of the literature review.

Table 15.2 shows a more detailed analysis for the curve based on a nonlinear regression. The overall $p < .0059$ was obtained which shows statistical significance of the polynomial model. The form of this relationship is:

$$\text{Success Measure} = B_0 + B_1 * \text{Planning effort index} + B_2 * (\text{Planning effort index})^2$$

The fit of this relationship is, however, low with R^2 less than .01. This suggests a small causal relationship indicating that less than 1% of project success can be attributable to the amount of effort spent planning. This is counterintuitive and I thought it deserved further analysis. A number of variables were examined to find the nature of their impact on the relationship between planning and success.

IMPACT OF MODERATORS

I first tried to find which of the moderators I identified were important to this relationship. A detailed series of regression analyses were carried out to analyze the impact of each moderator on the relationship. The results appear in Table 15.3. The majority of the moderators did not have a significant impact on the relationship between planning and success. The variables found to be moderators or quasi-moderators were included in the model (Experience level of team, Internal versus vendor based, Quality of WBS).

To complete the analysis, I first completed a regression analysis using the interaction terms from the moderator and quasi-moderator terms. When I completed a Moderated Hierarchical Regression Analysis (MHRA), using these interaction relationships, we get the results shown in Table 15.4. MHRA analysis is used to discover the underlying relationship between dependent and independent variables and understand how it is affected by moderating variables per Sharma, Durand, and Gur-Arie (1981). MHRA analysis enables us to explore these relationships in detail.

We can see that through the moderator analysis a more significant relationship between planning effort and project success has been uncovered. At R^2 = .14, we have a significant relationship between planning effort and success with $p < .001$.

If we complete a general regression analysis with only the interaction terms, we see the results shown in Table 15.5. The result of this model is

TABLE 15.3

Summary of Moderator Findings for Dependent Variable Success

Moderator	Role versus Project Success
Industry	Potential predictor
Geographic location	Predictor
Local versus international projects	Predictor
Stakeholder engagement level	Independent variable
Applicability/quality of the goals/vision	Independent variable
Quality of WBS	Independent variable and moderator
Methodology type (traditional vs. agile)	Independent variable and potential moderator
Novelty to organization	Independent variable
Technology level of the project	No relationship
Project length	No relationship
Project complexity	No relationship
New product versus maintenance	No relationship
Experience level of team	Independent variable and moderator
Internal versus vendor-based	Moderator
Team size	No relationship

TABLE 15.4

MHRA Analysis for Team Size as Moderator in the Planning Effort
Index versus Success Measure Relationship

Variables Entered	Step 1	Step 2	Step 3
Main Effects			
Planning effort index	1.972[b]	2.030[b]	13.007[c]
Planning effort index[b]2	−4.044[b]	−4.103[b]	−25.064[c]
Moderators			
Internal versus vendor based		.028	.010
Interaction Terms			
WBS[a]Planning effort index			−2.927[c]
WBS[a]Planning effort index[b]2			4.662[c]
Experience[a]Planning effort index			−3.965[c]
Experience[a]Planning effort index[b]2			8.944[c]
Internal[a]Planning effort index			.619[d]
Internal[a]Planning effort index[b]2			−1.330[d]
R^2	*.006*	*.016*	*.144*

[a] $p < .05$.
[b] $p < .01$.
[c] $p < .001$.
[d] $p < .10$.

TABLE 15.5

Multiple Regression of Final Model against Success Measure
with Moderator Interaction Terms

**Regression Summary for Dependent Variable:
Success Measure**

$R = .387$ Adjusted $R^2 = .145$ $p < .0001$

	B	p-Level
Intercept	3.222	0.000
Planning effort index	12.648	0.000
Planning effort index^2	−24.405	0.000
WBS Planning effort index	−2.924	0.000
WBS Planning effort index^2	4.653	0.000
Experience Planning effort index	−3.960	0.000
Experience Planning effort index^2	8.928	0.000
Internal Planning effort index	0.713	0.000
Internal Planning effort index^2	−1.495	0.005

TABLE 15.6

Summary of Main Findings of Binomial Relationship between Planning Indexes and Success Factors

	Base			Moderators Included		
	Overall Success Factor (R^2)	Success Factor (R^2)	Efficiency Factor (R^2)	Overall Success Factor (R^2)	Success Factor (R^2)	Efficiency Factor (R^2)
Planning effort index	.006[b]	.006[b]	.003[a]	.145[c]	.142[c]	.079[c]

[a] $p < .05$.
[b] $p < .01$.
[c] $p < .001$.

both a very good p-value $< .0001$ and a relatively strong $R^2 = .145$. Tests of residuals for this model also showed good results and confirmed the normality of the data. Normal probability plots, p–p plots, and homoscedasticity plots are included in Appendix B, Figures B.1–B.9.

This model was also regressed against the stakeholder success measure and efficiency measure. The stakeholder success measure produced very similar results, whereas regression against the efficiency measure had a good p-value but a lower R^2 of .079. See Table 15.6.

THE FINAL MODEL

To summarize, this final model confirms the relationship between planning effort and success with both low p and a substantial R^2. We can therefore report a significant link between project planning effort and project success. In addition, a model has been developed that is both empirically and logically consistent. This model shows a significant relationship between planning effort and project success. Sensitivity analysis of the moderating variables confirms the consistency and robustness of the model.

The approach of most previous researchers is partially validated by this research. The most important aspect of the planning phase is the quality. A relationship has been found in my research between time spent planning and success, however, that relationship has a R^2 value of .15, where measures of quality from the literature report effects of $R^2 = .34$ on average. It is the quality of those planning deliverables that is most important and that is where most previous research focused (Pinto and Prescott, 1990; Deephouse et al., 1995; Dvir et al., 2003; Zwikael and Globerson, 2006; Gibson et al., 2006).

Although the amount of effort spent is a somewhat crude measure of planning, it does tell us something about the project management process. The R and R^2 values found tell us that increasing the time spent planning will not guarantee project success. It is still significant that there is an amount of planning phase effort that maximizes project success. This relationship cannot be dismissed as it documents the situation in the available data with good statistical significance. The values found may be a product of the varied nature of the data and projects but are significant across a wide range of global projects and industries. In reviewing the scatter diagram (Figure 15.2), it is clear that the pattern is so complex, a solution would not be apparent without the computer analysis tools we have available today.

To summarize, we can confirm

There is a relationship between planning effort and project success: a model was derived that showed that this relationship had an R^2 of .15.

16

What Is the Optimum Planning Phase Time?

Without leaps of imagination or dreaming, we lose the excitement of possibilities. Dreaming, after all is a form of planning.

Gloria Steinem

OPTIMAL PLANNING EFFORT

I then wanted to take my research a bit further. Was there any useful guidance I could give to project managers, program managers, or executives? If we look at the curve for the relationship between planning effort and success, I thought there should be a maximum at the peak. Because there is a quadratic relationship between planning effort and success measures, it was possible to calculate a maximum to the resulting quadratic curve. If you remember from high school, for a quadratic equation of the form $Ax^2 + Bx + C$, the formula for the maximum is $B/2A$. After completing this calculation for the overall success measure, this value was found to be 0.249 or approximately 25% of effort spent on planning prior to execution. This is a very interesting result that indicates, on average, projects which spend approximately 25% of their effort on the planning phase are the most successful.

We continue the analysis with the narrower success measures as shown in Figure 16.1. In this case, the narrower success measure also showed a correlation with project planning with a very similar curve to the one seen with the analysis against the overall success measure.

We can also continue with the same analysis using the efficiency measure, Figure 16.2.

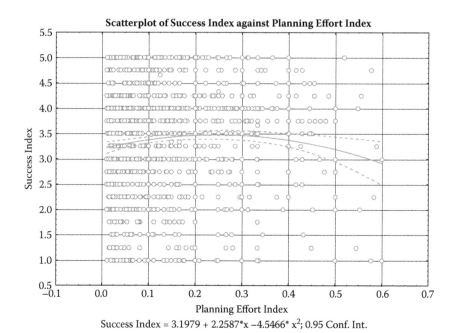

Success Index = 3.1979 + 2.2587*x −4.5466* x^2; 0.95 Conf. Int.

FIGURE 16.1

Scatterplot and curve fitting for success measure versus planning effort index.

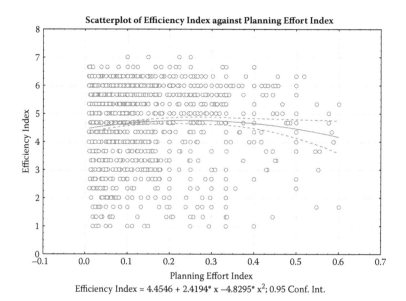

Efficiency Index = 4.4546 + 2.4194* x −4.8295* x^2; 0.95 Conf. Int.

FIGURE 16.2

Scatterplot and curve fitting for efficiency measure versus planning effort index.

A similar relationship is seen between the planning effort index and the efficiency measure. Table 16.1 gives more detailed statistical information for the curve based on a nonlinear regression. However, in this case, a *p*-value of .04053 was obtained, showing this relationship is not as statistically strong as the overall success measure but is still statistically significant.

We now calculate the planning effort index value that maximized the stakeholder success measure and efficiency measure in the same way we did for the success measure. Table 16.2 summarizes those calculations.

These results are interesting from a number of viewpoints. They are in line with the approximately 20–33% effort spent on planning identified in the literature review (Wideman, 2000; Nobelius and Trygg, 2002; Chatzoglou and Macaulay, 1996). Second, this result is lower than the $R^2 = .33$ correlation with efficiency and $R^2 = .34$ with success reported from the literature review meta-analysis (Table 16.3). This implies that there can still be a return on investment from spending 25% of effort on the planning phase. The three results are also within .01 of each other, which helps to validate the research methodology. Finally, they are higher than the average planning efforts reported by survey respondents.

TABLE 16.1

Nonlinear Regression Analysis of Planning Effort Index versus Efficiency Measure

Nonlinear Regression Summary for Planning Effort Index versus Efficiency Measure			
	B_x	R^2	*p*-Level
Intercept	4.46		0.00
Planning effort index	2.42		0.01
Planning effort index^2	−4.83		0.02
Overall Model		.003	.041

TABLE 16.2

Optimum Planning Effort Index Values by Success Measures

	Success Measure	Stakeholder Success Measure	Efficiency Measure	Mean	Actual Reported Planning Index Average
Planning effort index	0.255	0.248	0.250	0.251	.153

TABLE 16.3

Summary of Subgroup Analysis for Optimum Planning Levels

	p	Number of Projects	Average Planning Effort	Optimum Planning Level
Region – North America	0.031	756	0.151	0.226
Team type – International	0.03	442	0.149	0.216
Industry – Professional services	0.087	54	0.139	0.251
Industry – Education	0.118	42	0.132	0.210
Industry – Government	0.105	152	0.126	0.147
Industry – Retail	0.108	30	0.173	0.509

MAXIMUMS FOR SUBSETS

We can see from Table 16.2 that the optimum planning amounts are relatively consistent between the three success measures. In addition, optimum planning values were calculated on subsets of the data. The results were often within the .20–.25 range.

Subgroups were also analyzed to confirm optimum planning levels. Table 16.3 is a summary of some of those analyses. The statistical significance level ($p < = .05$) was relaxed to $p \sim .10$ to allow a broader view. Note that these smaller groups do not have the very good p-values of the other parts of this book. Subgroups not noted in the table had even higher p-values and were not included. This is because several hundred projects are likely required for a statistically valid sample and in most cases this volume of data was not available for those subgroups.

The results show some of the variation between industries. Retail, for example, shows a very high optimum planning perhaps because for retail projects, the majority of work really is in the planning. There is little to build in execution compared to construction and IT, for example. This is an interesting result with only 30 datapoints. In general, however, the subgroup analysis is in the range of and validates the overall results.

This indicates that the time required for the planning phase is most dependent on project characteristics. This is as one would expect. Certain projects may require a long requirements gathering and analysis period; others may not. This view is also validated by the work of Choma and Bhat (2010): "Thus, the time spent in FEL does not determine the quality of planning; rather, it is the deliverables completed in FEL that are most correlated with results" (p. 7). FEL is defined as planning phase front-end loading.

The mean project planning effort reported by respondents was substantially lower than these values at .15 of effort. This confirms a view that should not be surprising to practicing project managers; not enough planning is being done and that if longer planning phases were the norm, there would be higher overall project success.

FURTHER ANALYSIS

I next wanted to see how the final model looked graphically and how we could visualize the impact of planning effort on success.

I started by graphing the curve produced by the final model. However, because there are three moderators affecting the shape of this curve, I needed to make assumptions on their values. Figure 16.3 shows the curve representing average values with the exception of internal versus external teams. Because for external teams, planning is largely tracked elsewhere, the first graph focuses on internal teams with averages for all other measures. We can see that this curve is not as shallow as the curve for the

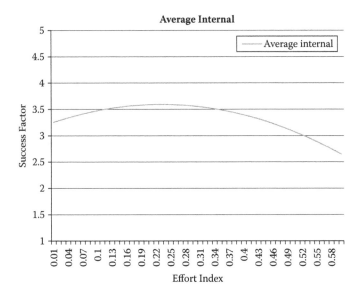

FIGURE 16.3
Planning effort index versus success for average moderator values for internal projects.

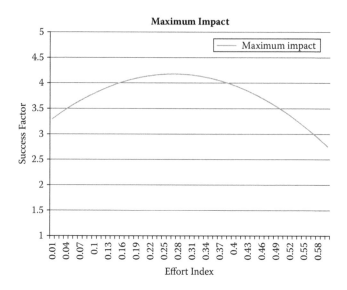

FIGURE 16.4

Planning effort index versus success for maximum moderator values.

whole dataset. This is to be expected because the moderated relationship had a higher R^2 value.

The graph in Figure 16.4 shows the impact of the planning effort index where all moderators are set for maximum impact. The graph shows an even steeper curve than the previous graph. In this case we can see that the impact is at its maximum with a project potentially affected by 1.5 success rating levels.

One can note from these graphs that the *y*-intercept indicates that for zero up-front planning projects still show average success. However, these curves refer to averages of variable projects with numerous moderators in play. Many projects also do substantial planning during execution.

CONCLUSION

The point we must take away is that, for the average project, doing no up-front planning will reduce its success rating by .5 to 1 success levels (i.e., turn a successful project to an average project, or an average project into an unsuccessful one). The impact on projects that end up in an overly long planning phase is even more severe at up to 1.5 success levels.

The final model clearly shows the importance of planning effort on success and shows that not applying the right effort can have an important impact on the success of projects.

The mean project planning effort reported by respondents at 15% of effort was substantially lower than the optimum planning effort values. Again, this confirms the view that not enough planning is being done and that if longer planning phases were the norm, there would be higher overall project success.

Therefore we can confirm

The optimum plan effort to maximize success is approximately 25% of total effort. Projects are planning substantially less than this on average.

17

Planning Budgets and Success

Economy does not lie in sparing money, but in spending it wisely.

Thomas Huxley

In some cases projects are primarily tracked by budget rather than effort. I also looked at planning budgets and their impact on success.

NEW RESEARCH ON PLANNING BUDGETS

This set of data suffered from lower rates of user response. This was to be expected; some projects do not track budgets closely. This is often the case in internal projects, where effort is tracked but because the customers are internal, budget is not tracked. In other cases, budget numbers were not managed by the project manager but were confidential at the senior account level. These issues were reported in the comments sections of the survey by some respondents. Therefore, for analysis of agile effort or budget effort, only the subset of the valid 1,386 cases which also had valid agile or budget data was used. There were 1,037 projects that had adequate data for analysis.

The two key sets of data gathered were percentage or fraction of time spent planning and project success measures. Similar to the effort analysis, to facilitate the analysis of the planning budget as a percentage of the total, the following index was calculated:

$$\text{Planning Budget Index} = \frac{\text{total budget expended on the planning phase}}{\text{The total project budget}}$$

For consistency and simplicity, similar exclusion rules are used for budget analyses: cases with their respective indexes <.01 or >.6 were excluded. Projects with valid planning budget indexes totaled 1,037.

A similar analysis to the effort analysis was completed for the budget effort index. First normality was checked for the planning budget index. Both skewness and kurtosis were higher than for the planning effort index but are still acceptable (Field, 2009). See Appendix B for more detail on this. We can continue with the scatterplot analysis followed by regression analysis. A typical scatterplot is given in Figure 17.1.

Again, there is a good fit between the polynomial curve and the data. We can get more information if we complete a nonlinear regression analysis. Completing this analysis for the planning budget index versus the success measure, does show statistical significance with $p < .01$, as shown in Table 17.1.

If we look at the Stakeholder Success Measure only, we get Table 17.2. The results show a p-value $< .067$, which was not statistically significant for the purposes of my research. However, it is close to $p < .05$ so for this

All Success Factor = $3.1872 + 2.1198^* x - 4.9729^* x^2$; 0.95 Conf. Int.

FIGURE 17.1

Scatterplot and curve fitting for success measure versus planning budget index.

TABLE 17.1

Nonlinear Regression Analysis of Planning Budget Index versus Success Measure

Nonlinear Regression Summary for Planning Budget Index versus Success Measure			
	B_x	R^2	*p*-Level
Intercept	3.187		.000
Planning effort index	2.120		.004
Planning effort index^2	−4.973		.003
Overall model		*.006*	**.013**

TABLE 17.2

Nonlinear Regression Analysis of Planning Budget Index versus Stakeholder Success Measure

Nonlinear Regression Summary for Planning Budget Index versus Stakeholder Success Measure			
	B_x	R^2	*p*-Level
Intercept	3.222		.000
Planning effort index	1.916		.023
Planning effort index^2	−4.401		.022
Overall model		**.003**	**.067**

TABLE 17.3

Nonlinear Regression Analysis of Planning Budget Index versus Efficiency Measure

Nonlinear Regression Summary for Planning Budget Index versus Efficiency Measure			
	B_x	R^2	*p*-Level
Intercept	4.386		.000
Planning effort index	3.534		.001
Planning effort index^2	−8.284		.001
Overall model		*.009*	**.004**

case, I stretch the acceptable significance to $p < .10$ to allow some further analysis.

The results are better when we look at the efficiency measure in Table 17.3. Here $p < .004$ in this case which is better than $p < .01$. Therefore, overall, we can confirm that the planning phase budget affects project success although efficiency is affected more than the overall success.

OPTIMAL PLANNING BUDGETS

We now complete a similar exercise to calculate the planning index that maximizes the success measures. You may recall in Chapter 16, we calculated the effort that would maximize project success. We can do the same for the budget index, as shown in Table 17.4.

The results are very consistent with a difference of only .005 between them. The average value of the planning budget that maximized project success is 21.5% of the overall project budget. This is somewhat lower than the value I found from studying the effort values, but this is to be expected. Some project budgets consist of both staff hours and equipment purchases (Wideman, 2000). The planning phase occurs before execution and therefore costs are typically staff costs only (Pinto and Prescott, 1988; Hamilton and Gibson, 1996). Therefore the percentage of budget in the planning phase would be lower.

We can see from Table 17.5 that the optimum planning amounts are also consistent between the three success measures. In addition, optimum planning values were calculated on subsets of the data. The results were consistently within the .20–.23 range.

TABLE 17.4

Optimum Planning Effort Index Values by Success Measures

Measure	Planning Budget Index Value Maximizing Success Measure
Success Measure	0.213
Efficiency Measure	0.213
Stakeholder Success Measure	0.218
Mean	0.215

TABLE 17.5

Main Findings of Calculated Optimum Planning Index Values for Relationship between Planning Indexes and Success Measures

	Success Measure	Stakeholder Success Measure	Efficiency Measure	Mean	Actual Reported Planning Index Average
Planning effort index	0.255	0.248	0.250	0.251	.153
Planning budget index	0.213	0.218	0.213	0.215	.127

The mean project planning effort reported by respondents was substantially lower than these values at .127 of budget. This again confirms the view that not enough planning is being done and that if more planning were the norm, there would be higher overall project success. From this research, the indication is that more money should be spent on planning than the current average.

Therefore we can confirm

The optimum planning phase budget to maximize success is approximately 21% of total budget. Projects are planning substantially less than this on average.

18

Planning and Agile/Iterative Methods

It is a bad plan that admits of no modification.

Publilius Syrus (~100 BCE)

Of course, not all planning is done in the planning phase. Some planning is integrated within the execution phase of projects and has been for a long time. For agile projects, this is a major tenet of the methodology. Planning during execution may consist of a substantial amount of the planning. All plans need to expect changes during project execution; however, the amount of planning done during execution for agile projects is usually more substantial than for typical projects. It is definitely more structured.

Koontz noted as early as 1958 that "[N]o effective manager makes a plan and then proceeds to put it into effect regardless of what events occur" (p. 54). Agile methods use a minimum of documentation to facilitate flexibility and responsiveness to changing conditions. This might imply that less planning is used in agile projects than in traditional project management.

AGILE

Agile methods have become more and more common in technology projects since their development (Lindvall et al., 2002). Other iterative methodologies, such as rolling wave (Turner and Cochrane, 1993), have been in use for years and can be thought of as predecessors to agile methods. As part of their rationale for the use of rolling wave, Turner and Cochrane (1993) noted, "[F]rozen objectives become part of the definition of the quality of the project, and project managers are said to be successful if they deliver them on time and within budget, regardless of whether or not the product is useful or beneficial to the owners and users" (p. 94). This highlights the benefits of iterative planning, which allows the replanning of a project during execution. Even before the advent of agile, it was known that 50% of design activities occurred in phases other than design (Fitzgerald, 1996).

Deviations from the original plan are common. Hällgren and Maaninen-Olsson (2005) define a deviation as "A deviation is, therefore, recognized as a situation, regardless of consequence—positive or negative, large or small—that deviates from any plan in the project" (p. 18). Deviations from plan will inevitably occur regardless of how well a project is planned or executed and although planning and change control are important to a project, neither solve deviations on their own. The solution is in the methodologies that should facilitate the necessary resolutions.

The following areas, for example, often change during project execution (Collyer et al., 2010, p. 113):

- Changing materials, resources, tools, and techniques
- Changing relationships with other related projects, services, or products
- Changing goals

Collyer et al. (2010) advocate a strategy of aim, fire, aim: that is, plan and execute but replan during the project to adapt to changes. This is similar to the rolling wave processes (Turner and Cochrane, 1993). They highlight the drawbacks of only having a comprehensive planning phase at the start of a project and state it is more effective to have waves of planning scheduled within execution. This allows the project to react to setbacks and opportunities that occur during the project and allow replanning based on these occurrences.

Of course, planning only at the start of a project can have negative consequences. For example, Boehm (1996) discussed challenges in software development. He noted that a too-detailed requirements document can have problems (p. 74):

1. Specifications do not describe a deliverable as well as a prototype.
2. Having to specify requirements in advance results in gold-plating (adding more features than required) as there will be no further opportunities to add functionality.
3. The solutions are focused on a point in time and the requirements or environment may then change.

However, he notes that evolutionary or iterative models also have faults:

1. The initial increment is not designed adequately for growth.
2. Early partial prototypes set unrealistic expectations; the hard work of behind the scenes coding and performance tasks are left until later.
3. Initial releases may have such limited capabilities that users fail to bother to learn or use the system.

It is therefore important to plan and execute but it is also important to replan during the project to adapt to changes.

UP-FRONT PLANNING IN AGILE/ITERATIVE

There are clear drawbacks of only having a comprehensive planning phase at the start of a project and this has been known for a long time. It is more effective to have additional planning scheduled within execution, whether they are called waves, spirals, iterations, or story planning sessions. These kinds of planned review activities allow the project to react to setbacks and opportunities that occur during the project and allow replanning based on these occurrences.

However, forging forward with no up-front planning has its drawbacks. I once worked for a small start-up company with grand ambitions. They had partnerships in place with major players and were in what seemed to be a growth area. I was brought on to help manage implementation of the systems strategy put in place by the CIO and president. After a couple of months, the CIO went on extended leave. At this point, I took on a greater role in the program and started to examine the strategy. A dual-site, redundant system was being built even though the hosting company already had redundant backups. I asked a question, "Had any clients requested this?" If not, then why were they spending a huge amount of money and time to build this? No one had an answer. Assumptions were made without enough analysis. In the end, the second site was removed as unnecessary gold-plating but not before time and money had been spent unnecessarily. Questioning and then replanning allowed me to save this company substantial money.

Collyer and Warren (2009) note that it is important to have flexibility and be able to adapt: "Pfizer's disappointing heart medication,

Viagra, turned into a success because they took the time to investigate its side effects" (p. 359). Not keeping to the original plan paid handsome dividends; it became one of Pfizer's most profitable drugs. This kind of opportunity cannot be foreseen at the start of a project but must be reacted to during execution. This can apply to many other fields and not just pharmaceuticals or even new product development. Taking time to replan or overcome problems or seize opportunities is crucial in a variety of industries.

Boehm reported that by using a spiral model in IT development, which included a planning phase and execution phase in each spiral, "A cost improvement from $140 to $57 per delivered line of code and a quality improvement from more than 3 to 0.35 errors per thousand delivered lines of code." A spiral is a phase similar to a wave in rolling wave or stories in agile. These examples show the benefits of replanning and flexibility in project management.

Given the reported failure rates of IT projects (Sessions, 2009; Standish Group, 2011), it was felt that a formal method of managing requirements changes and evolving IT projects was required. The agile movement was intended to address some of the challenges of poor performance due to having one rigid project plan.

> In 2001, the "Agile Manifesto" was written by the practitioners who proposed many of the agile development methods. The manifesto states that agile development should focus on four core values:
>
> - Individuals and interactions over processes and tools.
> - Working software over comprehensive documentation.
> - Customer collaboration over contract negotiation.
> - Responding to changeover following a plan.
>
> Dybå and Dingsøyr (2008) and www.agilemanifesto.org.

Dybå and Dingsøyr (2008) go on to define the difference between agile and tradition software development methodologies as summarized in Table 18.1. Some of these principles are gaining growing acceptance in the software industry in particular and are making inroads in the project management community in general (Dybå and Dingsøyr, 2008; Coram and Bohner, 2005). Agile has further developed since its inception to include a number of difference methodologies that encompass the agile philosophy. For example, there are a number of agile software development methods as noted in Table 18.2.

TABLE 18.1

Main Differences between Traditional Development and Agile Development

	Traditional Development	**Agile Development**
Fundamental assumption	Systems are fully specifiable, predictable, and are built through meticulous and extensive planning	High-quality adaptive software is developed by small teams using the principles of continuous design improvement and testing based on rapid feedback and change
Management style	Command and control	Leadership and collaboration
Knowledge management	Explicit	Tacit
Communication	Formal	Informal
Development model	Life cycle model	The evolutionary-delivery model
Desired organizational form/structure	Mechanistic (bureaucratic with high formalization), aimed at large organizations	Organic (flexible and participative encouraging cooperative social action), aimed at small and medium-sized organizations
Quality control	Heavy planning and strict control	Continuous control of requirements, design, and solutions
	Late, heavy testing	Continuous testing

Source: After T. Dybå and T. Dingsøyr, *Information and Software Technology* 50: 833–859, 2008.

PLANNING IN AGILE

But what are the planning needs of agile methodologies? Smits (2006), in a white paper on planning in agile methods, notes that there are different types of planning that are important. He identifies five levels of planning for agile projects:

Product Visioning – Level 1
Product Road Map – Level 2
Release Planning – Level 3
Iteration Planning – Level 4
Daily Plan – Level 5

He describes the need for the higher-level planning: "The experienced disadvantage of iteration planning when applied to projects that run for more than a few iterations or with multiple teams is that the view of the longer term implications of iteration activities can be lost" (p. 4). He also notes that substantial planning is completed in daily meetings: "This daily meeting is not often seen as a planning session, but certainly is" (p. 8).

TABLE 18.2

Description of Main Agile Development Methods

Agile Method	Description
Crystal methodologies	A family of methods for colocated teams of different sizes and criticality: Clear, Yellow, Orange, Red, Blue. The most agile method, Crystal Clear, focuses on communication in small teams developing software that is not lifecritical. Clear development has seven characteristics: frequent delivery, reflective improvement, osmotic communication, personal safety, focus, easy access to expert users, and requirements for the technical environment.
Dynamic software development method (DSDM)	Divides projects in three phases: preproject, project life cycle, and post project. Nine principles underlie DSDM: user involvement, empowering the project team, frequent delivery, addressing current business needs, iterative and incremental development, allow for reversing changes, high-level scope being fixed before project starts, testing throughout the life cycle, and efficient and effective communication.
Feature-driven development	Combines model-driven and agile development with emphasis on initial object model, division of work in features, and iterative design for each feature. Claims to be suitable for the development of critical systems. An iteration of a feature consists of two phases: design and development.
Lean software development	An adaptation of principles from lean production and, in particular, the Toyota production system to software development. Consists of seven principles: eliminate waste, amplify learning, decide as late as possible, deliver as fast as possible, empower the team, build integrity, and see the whole.
Scrum	Focuses on project management in situations where it is difficult to plan ahead, with mechanisms for "empirical process control"; where feedback loops constitute the core element. Software is developed by a self-organizing team in increments (called "sprints"), starting with planning and ending with a review. Features to be implemented in the system are registered in a backlog. Then, the product owner decides which backlog items should be developed in the following sprint. Team members coordinate their work in a daily stand-up meeting. One team member, the scrum master, is in charge of solving problems that stop the team from working effectively.
Extreme programming (XP; XP2)	Focuses on best practice for development. Consists of 12 practices: the planning game, small releases, metaphor, simple design, testing, refactoring, pair programming, collective ownership, continuous integration, 40-hour week, on-site customers, and coding standards. The revised "XP2" consists of the following "primary practices": sit together, whole team, informative workspace, energized work, pair programming, stories, weekly cycle, quarterly cycle, slack, 10-minute build, continuous integration, test-first programming, and incremental design. There are also 11 "corollary practices."

Source: After T. Dybå and T. Dingsøyr, *Information and Software Technology* 50: 833–859, 2008.

Similarly, Coram and Bohner (2005) note that agile methods do require up-front planning. Working with the customer is needed to provide requirements for the first release. They also note, "With so many small tasks, it is argued that agile processes require more planning. … [I]t is a constant task to ensure optimal delivery results" (p. 6). Agile methodologies forgo a long planning phase but break planning activities into iterations more closely tied with execution and testing. This can take a few weeks to a few months.

But evolution of systems is not always the best strategy. Good up-front planning and analysis can also be critical. I worked for a company that experienced unexpected high growth for many years. Their databases were originally designed for 10,000 customers but were asked to contain records for 4 million people. Unfortunately, the original designers did not have the foresight to design an easily expansible architecture. This became worse as more and more modules were bolted on to this shaky structure. But the business sponsors would rather spend their budget on projects that would allow them to keep growing rather than projects to clean up and redesign clunky systems. On top of this, the business could not seem to understand why the systems were so hard to test, seemed to have a lot of bugs, or why projects were consistently late. Yet they were unwilling to fund the necessary changes. The relationship between business and IT became a nightmare. One can argue that this is the type of trap agile projects can fall into. Without a detailed up-front planning and analysis phase, sizing decisions are not made properly and long-term goals and directions are not properly accounted for in design.

Boehm (2002) in reviewing agile methods and comparing them to traditional methodologies notes both sides of the issue. When projects have excessively specified plans: "Such plans also provide a source of major contention, rework, and delay at high-change levels" (p. 65). However, a balance between traditional planning and agile methods is usually appropriate. Certain factors, such as the size of the project, safety requirements, and known future requirements, call for up-front planning even in agile projects, whereas turbulent, high-change environments call for less up-front planning and a greater use of agile methods. He notes there is a "sweet spot" that is dependent on project characteristics where the effort expended in initial planning pays off in project success. Too much or too detailed planning can result in wasted effort and too much plan rework, whereas not enough initial planning can result in project failure. He does not give guidance on whether there are methods to find this "sweet spot" but notes its existence. He also notes that even pure agile methods represent a significantly higher degree of planning than undisciplined hacking (p. 64).

Mann and Maurer (2005) found, in a study on the impact of Scrum (an agile methodology) on overtime and customer satisfaction, that customers believed that the daily meetings kept them up to date and that planning meetings were helpful to "reduce the confusion about what should be developed" (p. 9). As well, they report:

> One customer had this to say about the planning meetings, "Superb forum for planning; the whole team is involved and thus everyone knows what is required from them." Another customer mentioned how they think that the planning meetings prevent problems later on: "Although the day as a whole can be a very tiring process, I have found that the time spent in the planning meetings has led to less misdirected development and a more clear understanding of both the requirements and the limitations of the development process by both the customers and the developers." (p. 8)

One could be led to believe from these quotes that the structure of agile planning is one of the reasons that agile is successful. Dybå and Dingsøyr (2008) reported that the planning game activity, which is one of the techniques for planning used in agile methodologies, was found to have a positive impact both in the company and with customers because it gave both groups insight into the development process. Koskela and Abrahamsson (2004) analyzed the role of the customer in an XP project (or extreme programming, which is another agile methodology) and found that most of the time was spent on participating in planning game sessions and acceptance testing, followed by retrospective sessions at the end of release cycles. This highlights the important of some planning in even agile projects which strive to reduce the amount of formal process that is used. In fact, planning took up 42.8% of customer's effort which shows the importance of planning in agile environments. See Figure 18.1.

I have personally managed extreme programming projects and can attest they do produce results. In one case, the project was delivered on time and on budget with very few bumps. However, the budget was a bit higher that I would have expected. So there may be some costs to having pairs of programmers rather than individuals. As I find in Chapter 19, agile can lead to more consistently successful projects but not necessarily consistently cheaper ones.

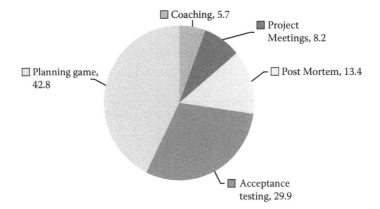

FIGURE 18.1
Customer effort distribution (%) for XP projects. (After J. Koskela and P. Abrahamsson. In T. Dingsøyr (Ed.), *Software Process Improvement,* Vol. 3281 (pp. 1–11), 2004, New York: Springer.)

AGILE VERSUS TRADITIONAL

Magazinius and Feldt (2011) examined the variation in planning between agile and nonagile companies, that is, companies that have or have not adopted agile methodologies. They reported that the success in meeting time and budget goals and the causes of failures were not significantly different between the two methodologies. They also suggest that although estimation techniques have improved over time, factors such as the overly aggressive push to reduce costs are at work to distort estimates. This can result in adverse outcomes such as incorrect projects being selected, over-runs costing more than the original rejected estimate or projects that might have been more beneficial being passed over.

Ceschi and colleagues (2005) studied a data sample comprising managers of software companies: 10 adopting agile methods and 10 using traditional approaches. They state, "Most agile companies tend to plan and, consequently, to develop only essential functionality at each iteration. However, this doesn't mean they don't carefully plan their development processes. In fact, they're more satisfied with the way they plan their projects than plan-based companies are" (pp. 23–24). Interestingly they found that even in agile "Eighty-five percent of the managers would like to improve process planning, even though 70% of managers are sufficiently satisfied with it and 20% are very satisfied" (p. 24). Managers of agile projects were slightly more satisfied with their project planning: 20% of nonagile managers were dissatisfied with their planning, whereas none of the agile project managers was dissatisfied with their planning process.

From the literature we can therefore note the following:

Planning is required in interactive or agile methodologies, both up-front planning and during execution.

19

Agile Methods and Success

Just because something doesn't do what you planned it to do doesn't mean it's useless.

Thomas A. Edison

OVERVIEW OF AGILE METHODOLOGY AND SUCCESS RATES

As part of my original research, I also gathered information on the planning characteristics of agile and iterative projects. A similar research construct was used for agile as was used for the initial planning investigation. See Figure 19.1.

As you may remember, I measured success using three success measures and these same measures are used to study agile impact. These were:

Efficiency Measure
Stakeholder Success Measure
Success Measure

The two key sets of data gathered for the purpose of this survey were percentage or fraction of time spent planning and project success measures. For projects with agile/iterative planning components an index was created:

$$\text{Agile Planning Effort index} = \frac{\text{total effort expended on planning after the planning phase}}{\text{the total project effort } (\text{in person-days})}$$

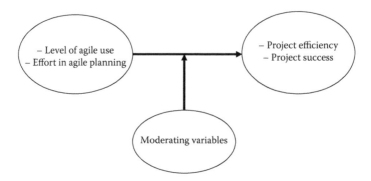

FIGURE 19.1
Main research constructs for agile.

The amount of planning done in the planning phase was considered a key dataset. The planning effort index was used as the key measure of this information (more details in Appendix A). After removal of invalid data and outliers, valid agile projects totaled 1,002.

Agile and iterative methods had a large impact on the projects. In my survey, I asked respondents how much agile or iterative methodologies they used in their project. More than 65% of the original 1,386 projects reported having some agile or iterative component. See Table 19.1.

You may wonder why I asked about the percentage of agile use in a project. Agile methods are not an all-or-nothing proposition. In my career I have used both full and partial agile methods. On one project, I was managing an offshore vendor. It was a new team working on a challenging application. I wanted to use agile methods to reduce risk because the development team was remote and because I couldn't manage them in person, I chose to use just some aspects of agile. I requested they add regular application builds to their

TABLE 19.1

Frequency Table for Methodology Type

Percentage Agile/Iterative	Count	Cumulative— Count	Percent
80–100%	80	80	5.8
60–79%	152	232	11.0
40–59%	347	579	25.0
20–39%	162	741	11.7
1–19%	194	935	14.0
0%	451	1,386	32.5
Missing	0	1,386	0.0

plans instead of only building the full application at the end of the project. In addition, I asked them to build demos into the plan. The results were predictable: the first integrated build took four weeks longer to complete than planned and the first demo didn't work at all. But by the end of the project, builds were running smoothly and the demos were showing what we needed to see. These demos also clearly provided useful insights to users. I was not able to use a fully agile methodology but I took some aspects of agile and used them to the benefit of the project.

The more of an agile/iterative approach that was reported, generally the higher was the reported project success. In Table 19.2 we can see from the *p*-value for stakeholder success measure and success measure that there is clearly a relationship between agile use and the three success measures. The amount of planning during execution is also significantly related to methodology as expected. We know from Chapter 18 that agile projects do substantial planning in the execution phase.

In addition, the correlation between the level of agile development and success is small but statistically significant as shown in Table 19.3 It is interesting to note that projects with a high agile percentage report mean up-front planning amounts similar to traditional projects and there is no relationship between agile use and upfront planning as shown by the *p* for the upfront planning index. If substantial planning is done during execution as noted by Dybå and Dingsøyr (2008) and Smits (2006), then agile projects appear to do more planning overall than traditional projects.

TABLE 19.2

Methodology Type Means: How Much of the Project Was Done Using Agile or Iterative Techniques

	Means and ANOVA					
Percentage Agile/ Iterative	**Upfront Planning Effort Index**	**Agile Planning Effort Index**	**Success Measure**	**Efficiency Measure**	**Stakeholder Success Measure**	**Valid *N***
80–100%	0.16	0.15	3.57	4.82	3.64	80
60–79%	0.15	0.14	3.48	4.66	3.57	152
40–59%	0.16	0.13	3.50	4.79	3.54	347
20–39%	0.14	0.10	3.36	4.64	3.41	162
1–19%	0.15	0.09	3.17	4.46	3.18	194
0%	0.15	0.05	3.22	4.58	3.21	451
All Groups	0.15	0.11	3.38	4.65	3.38	1,386
$p(F)$	0.17	0.00	0.00	0.09	0.00	

TABLE 19.3

Correlation Analysis between Methodology Type and Success Measures

	Project Success Rating	Efficiency Measure	Stakeholder Success Measure
Methodology type	.172	.062	.157
	p = .000	*p = .022*	*p = .000*

TABLE 19.4

Standard Linear Regression Analysis of Methodology Type versus Project Success Rating

Regression Summary for Methodology Type versus Project Success Rating				
	R	$R^2 =$	Number of Projects	*p*-Level
Methodology type	.172	.029	1,386	0.000

Next, I performed a similar analysis on methodology and various success measures, as shown in Table 19.2. What is indicated is that agile methodologies are correlated with higher reported success, both efficiency and overall project success; the highest correlation is with stakeholder success. This correlation is significant given the low *p*-values.

We can also perform a regression analysis to further understand the relationship, as shown in Table 19.4. We find a relationship with a very low *p*-value albeit with an also low R^2 of .03. MHRA could help further analyze this relationship and by removing the moderator impact, find the underlying relationship and a higher R^2.

These results are convincing as they show a very low *p*-value. However, some may still question the issues of basing results on single questions and in this case the amount of agile use was asked in a single question.

To avoid the issue of regression against a single item measure, a new index was created. This index combined the results of the question on the degree of agile use in the project with the measure of how much planning was done in the execution phase. The following factor was defined.

Combined Agile Factor = mean of the following two responses as a summated scale of normalized values:

1. Methodology type
2. Agile planning index

where agile planning index is defined as follows:

$$\text{Agile Planning Index} = \frac{\text{total effort expended on planning after the planning phase}}{\text{The total project effort (in person-days)}}$$

Both items are taken as a measure of the "agileness" of the project. The first is based on respondent assessment of how much agile process is used in the project, and the second on how much planning is completed during execution which is a feature of agile projects (Dybå and Dingsøyr, 2008; Smits, 2006).

The data I had on agile required some transformation prior to analysis due to the way the questions were answered. To facilitate the analysis from the data responses, I broke the agile data into two groups: Group 1, which had a tendency to plan more before execution and Group 2, which contained some projects that planned more during execution. See Tables 19.5 and 19.6.

One can see that for the second group, the p-level is even better although R^2 values are somewhat lower compared with the Group 1 result and with the single factor. All analyses, however, point to a case for the use of agile methods being positively associated with improved success.

TABLE 19.5

Standard Linear Regression Analysis of Combined Agile Factor versus Project Success Rating for Group 1

Regression Summary for Combined Agile Measure versus Project Success Rating				
	R	$R^2 =$	Number of Projects	p-Level
Combined Agile Measure	.133	.015	412	0.007

TABLE 19.6

Standard Linear Regression Analysis of Combined Agile Factor versus Project Success Rating for Group 2

Regression Summary for Combined Agile Measure versus Project Success Rating				
	R	$R^2 =$	Number of Projects	p-Level
Combined Agile Measure	.120	.013	590	0.003

AGILE USAGE AROUND THE WORLD

Is there variability in agile use among different geographic regions? I now examine the differences between regions. See Table 19.7.

It is interesting to note that there are no clear patterns when the projects are grouped by regions. Areas that report high use of agile (higher values in the methodology column) do not show higher success ratings. Also, there was no significant relationship found between agile use and success in any region other than Latin America. However, the overall relationship between agile use and success is found in the All Groups category which contains the full dataset. This appears to be a case where the smaller numbers of datapoints in each region made finding statistical significance difficult. We can note that agile is most common in Russia and the Pacific, followed by Europe, India, Middle East, and North America.

AGILE USAGE BY INDUSTRY

Next, a similar analysis was completed by industry. In this case, shown in Table 19.8, some interesting patterns could be seen. Four areas showed statistical significance: high technology, health care, professional services, and other. This is not surprising as agile is more widespread in the high tech and IT fields and in fact agile was originally designed for that type of environment (Dybå and Dingsøyr, 2008). Industries where you would not expect to find agile methodologies such as construction, manufacturing, and retail do not show a statistically significant relationship.

AGILE USAGE ANALYZED USING MODERATORS

If we do a MHRA analysis using the moderators collected during the study we have mixed results. For the Group 1 data no moderator could be found. However, for the Group 2 data, a strong moderator was found in the quality of goals/vision. This moderator likely indicates the importance of the quality and clarity of the goals given to the project team. See Tables 19.9 and 19.10. With the Group 2 data, the R^2 value reaches .17 when the effect

TABLE 19.7

Comparison of Means and Regression Results for Agile Success by Region

	Methodology Type	Planning Effort Index	Success Factor	Project Success Rating	Efficiency Factor	Stakeholder Success Factor	Valid N
Indian subcontinent	3.67	0.17	3.32	3.38	4.51	3.39	97
North America	4.20	0.15	3.44	3.42	4.79	3.47	756
Africa sub-Sahara	4.62	0.22	3.20	3.11	4.44	3.24	37
Australasia	4.33	0.16	3.22	3.35	4.45	3.22	49
Arctic and Antarctica	5.00	0.18	4.73	5.00	6.00	5.00	1
Europe	4.09	0.13	3.26	3.24	4.52	3.29	213
Latin America	4.37	0.16	3.07	3.10	4.23	3.10	83
Russia and FSU	3.75	0.16	3.25	3.42	4.17	3.42	12
Pacific	3.83	0.16	3.39	3.38	4.81	3.37	24
Middle East	3.85	0.19	3.25	3.23	4.43	3.32	82
Far East	4.44	0.16	3.14	2.88	4.77	3.00	32
All groups	4.20	0.17	3.39	3.41	4.65	3.44	1,386

TABLE 19.8

Comparison of Means and Regression Results for Agile Success by Industry

	Methodology Type	Efficiency Measure	Stakeholder Success Measure	Overall Success Measure	Valid N	Regression p-Value vs. All Success
Construction	4.7	4.5	3.7	3.5	23	0.2
Financial services	4.8	4.6	3.3	3.3	73	0.7
Utilities	4.3	4.4	3.4	3.3	23	0.6
Government	3.6	4.3	3.2	3.1	34	0.3
Education	3.8	4.9	3.2	3.3	10	0.1
Other	4.2	4.5	3.2	3.2	53	0.0002[a]
High technology	4.5	4.7	3.4	3.4	57	0.04[a]
Telecommunications	4.3	5.1	3.7	3.7	35	0.6
Manufacturing	4.8	4.4	3.3	3.3	42	0.7
Health care	4.5	5.0	3.5	3.5	24	0.02[a]
Professional services	4.3	4.5	3.3	3.3	22	0.03[a]
Retail	4.3	4.6	3.2	3.2	16	0.7
All groups	4.4	4.6	3.4	3.3	412	0.007[a]

[a] Indicates statistical significance.

TABLE 19.9

Summary of Moderator Findings for Dependent Variable Success in Group 2

Moderator	Role versus Project Success
Stakeholder engagement level	Independent variable
Applicability/quality of the goals/vision	Independent variable and moderator
Quality of WBS	Independent variable
Novelty to organization	No relationship
Technology level of the project	No relationship
Project length	Independent variable
Project complexity	Independent variable
New product versus maintenance	No relationship
Experience level of team	Independent variable
Internal versus vendor based	No relationship
Team size	No relationship

TABLE 19.10

MHRA Analysis for Quality of Goals/Vision as Moderator in the Agile Factor versus Success Measure Relationship for Group 2 Data

Variables Entered	Step 1	Step 2	Step 3
Main Effects			
Combined agile factor	.715[b]	.337	−1.444[a]
Moderators			
Quality of goals/vision		−.484[c]	−.825[c]
Interaction Terms			
Quality of goals/vision[a] Combined agile factor			.858[c]
F for regression	8.622[c]	57.476[c]	41.619[c]
R^2	0.013	0.161	0.171

[a] $p < .05$.
[b] $p < .01$.
[c] $p < .001$.

of the quality of goals/vision is taken into account with very good *p*. This value of R^2 is quite good for studies of project management and indicates a strong relationship between agile methodology and success.

Results are similar for the entire dataset. Table 19.11 is a summary of the MHRA analysis for the whole dataset and for the key success measures. For the whole dataset, the *p*-values are higher although still less

TABLE 19.11

Comparison for R^2 from MHRA Analysis
between Combined Agile Factor and Success
Measures for the Full Dataset

	$N = 1,002$		
	Stakeholder Success Measure	**Efficiency Measure**	**Success Measure**
R^2	.152	.096	.164
	$p = .089$	$p = .083$	$p = .070$

than $p < .10$. It is also interesting to note that the R^2 values are similar to the values for R^2 for Group 2: .164 versus 171. The use of agile methodologies also appears to have a greater impact on stakeholder success measures and overall success measures than straight efficiency. Of course, this is largely in keeping with one of the stated goals of agile methodologies: to focus on maximizing utility for the users or stakeholders.

CONCLUSION

The data analyzed by this research show a clear link between the use of agile methodologies and project success. This is in keeping with their growth in popularity over the last decade (Magazinius and Feldt, 2011).

Use of agile methodologies is correlated with success as shown in Table 19.4. As reported in that table, the greater the use of agile methods, the greater is the reported success. There is a statistically significant relationship based on the various analyses completed. This relationship was analyzed using a number of different methods. They consistently demonstrated a relationship between agile and success with an exception for certain industries and when the data subset became too small. The MHRA analysis showed that the relationship may have an R^2 as high as .17 which is a good result for a study of project management.

A great deal has been written in the press regarding the impact of agile processes on success; however, this is the first research to examine that relationship from an empirical standpoint with this volume of data and shows that it is, in fact, valid. This was validated both over a large dataset of 1,386 projects with a single measure and in a smaller dataset of 412 with a summated measure.

Agile methodologies do have a measurable impact on success and it is not surprising if they gain further acceptance over time in the project management community.

It is interesting to note that projects with a high agile percentage report mean up-front planning amounts similar to traditional projects. If substantial planning is done during execution, as noted by Dybå and Dingsøyr (2008), Coram and Bohner (2005), and Smits (2006), then agile projects appear to do more planning overall than traditional projects. The impact of the planning done during execution is another area for research. Does the relationship between how planning is structured affect agile success? Further research in this area is warranted.

There is a clear link between the use of agile methodologies and project success: the greater the use of agile methods, the greater is the reported success.

20

Planning's Importance to Manager Success

> The more time you spend contemplating what you should have done
> ... you lose valuable time planning what you can and will do.

Lil Wayne

According to Kerzner (2003), "The difference between the good project manager and the poor project manager is often described in one word: planning" (p. 426). Koontz (1958) states, "unless a manager's job includes at least some planning, there is doubt that he is truly a manager" (p. 50).

Does the amount of time spent in planning and analysis also affect an individual manager's success as well as project success? Crawford (2000) notes that planning ability is one of the top factors cited in the literature on project management competency (p. 7). It is therefore worth examining this area in the literature further. Boynton and Zmud (1984) also note that planning is a key senior management activity and that senior managers are more likely to be comfortable with planning concepts than more junior managers. Jiang, Klein, and Chen (2001) report that planning activities affect a project manager's performance; of course, that may be because planning is related to project performance. Successful projects no doubt help a project manager's performance.

MANAGER PLANNING TIME

However, there is a wider question: Rather than project or enterprise planning, is the time a manager spends planning his or her own work important for career success? Mintzberg (1975), in his work on the nature of a manager's job, makes the point that managers have very little time for planning and spend the majority of their time reacting and dealing with people: "[T]he job of managing does not breed reflective planners; managers respond to stimuli, they are conditioned by their jobs to prefer live to delayed action" (p. 165). However, he states that managers plan nonetheless, "When managers must plan, they seem to do so implicitly in the context of daily actions, not in some abstract process reserved for two weeks in the organization's mountain retreat" (p. 164). Crawford (2000) also notes that planning competency is associated with higher project management competency self-assessment. However, this link was not consistently found in the supervisors' assessment of project manager competency: project managers rate themselves better overall if they plan well; supervisors did not rate project managers who plan well better than their peers.

There were geographical differences shown in this study indicating that project managers' performance assessment varied between North America and Europe. UK managers tended to rate the performance of their project managers lower than did American managers. They also tended to rank UK project managers as poorer performers if the project managers had self-reported as having good planning knowledge or good quality knowledge. This trend was not seen in American or Australian managers who tended to agree somewhat more on the value of managers who rated themselves highly on planning skills. The author suggested it could be related to a British cultural emphasis on understatement and objectivity.

PLANNING STRATEGIES OF SUCCESSFUL MANAGERS

Kotter (1999) reports that effective general managers spend little time in formal planning. They appear to get more benefit by randomly talking about disjointed topics with lots of people who do not report to them. In his view formal or daily planning is not useful; managers obtain information continually, not just at planning meetings. However, effective general managers have overall strategies and plans in mind, even before they start a new role. He refers to these general strategies as agendas. He notes also that "Effective executives develop agendas that are made up of loosely connected goals and plans that address their long-, medium-, and short-term responsibilities" (p. 148).

Mintzberg (1994) notes that attempts by companies to remove the strategic planning function from managers and give it to a strategic planning department were generally not successful. He states strategic planners do not have the in-depth organizational knowledge or secondary information sources necessary to create an effective vision or strategy. It is important that managers perform their own strategic analysis and planning even if a strategic planning department provides analysis as an input or helps craft the final detailed plans.

Covey (2004) in his popular work on personal success notes the seven habits of effective people:

Habit One: Be Proactive
Habit Two: Begin with the End in Mind
Habit Three: Put First Things First
Habit Four: Think Win/Win
Habit Five: Seek First to Understand and Then to Be Understood
Habit Six: Synergize
Habit Seven: Sharpen the Saw

One can state that two of the habits identified (Begin with the End in Mind and Put First Things First) are really related to planning and three further of the habits (Think Win/Win, Seek First to Understand and Then to Be Understood, and Synergize) are related to analysis. One can argue this book is in some ways a call for more thoughtful planning and analysis as a path to personal success.

Buehler, Griffin, and Ross (1994) note that the so-called "Planning Fallacy" is particularly apt to affect people who are trying to estimate how much they personally can get done and when they can complete various tasks. This should imply that managers who spend more time planning should be able to get estimates and therefore results. The implication is that this type of estimating requires time and practice to get right.

Luthans (1988), who logged the behaviors and activities of 44 managers, noted that there is a difference in where effective versus successful managers spend their time. He found effective managers spent more time in traditional management activities such as planning, organizing, commanding, coordinating, and controlling whereas successful managers spend more time networking. He concludes,

The traditional assumption holds that promotions are based on performance. This is what the formal personnel policies say, this is what new management trainees are told and this is what every management textbook states should happen. On the other hand, more "hardened" (or perhaps more realistic) members and observers of real organizations (not textbook organizations or those featured in the latest best sellers or videotapes) have long suspected that social and political skills are the real key to getting ahead, to being successful. Our study lends support to the latter view. (p. 131)

He reports that managers who spend more time in tasks such as planning are more effective in producing results but not more successful in their careers. There are, of course, managers who are at the same time both more effective than average and more successful than average. Luthans' results are in agreement with some of the observations of Crawford (2000) who found that project manager's supervisors rated performance based on factors other than the traditional management competencies including planning.

In my career I have worked in situations that may bear this out. There are some project managers who become quite successful by appearing to their managers to be the ones saving the day. They work all hours and roll up their sleeves and get anything done that needs doing. This type of high-energy, hands-on manager can move quite far up the hierarchy. I remember working with one such program manager. She had an experienced, knowledgeable team of project managers working with her. However, if a crisis came up, she would be the one chairing the meetings, working weekends, and putting together tiger teams. The managers under her felt rather underutilized and some moved off the program partially out of boredom. Taking a step back, she could have delegated and focused on the program management role. It probably would have been more effective and more efficient. Did people like working with her? Was this good for the organization? Probably not, but her career did move forward.

Tullett (1996) studied the thinking style of managers by studying 203 managers of multiple projects. He compares the adaptive to the innovative thinking style as defined by Kirton (1976) and notes that managers who manage multiple projects tend to the innovative thinking style. He states in his conclusions,

It has been shown that a typical manager of multiple projects has an innovative thinking style. This means that s/he is likely to be less concerned with the attention to detail and the structured, systematic approach required to plan and manage such projects successfully. This is likely to be a contributing factor not only to budget overruns and the late delivery of projects but also to conflict with typically more adaptive project sponsors. (p. 286)

His implication is that choosing project managers who are more detail oriented (adaptive) and therefore better at planning would result in more successful projects and one would assume, better career results for the project managers. Crawford (2005) reported in a study of senior management perceptions of project manager competencies that planning was listed as number four out of sixteen factors. Having a detailed understanding of contracts and cost on the projects was rated one and two which one can argue is related to involvement in upfront and project planning activities. Senior management, in this case, rated the planning abilities of project managers as quite important which was not fully in agreement with the findings of Crawford in 2000. The importance of planning to manager success appears to have a variety of interpretations in the literature.

I recall one manager at a large corporation who had become a fixture in his role. He had been managing the same team successfully for years. He knew the application inside out and got along well with his employees. However, he also had a habit of complaining about his bosses and would have the occasional blowout at his manager. A few years later, I heard he had been cut in a reorganization. I can't imagine his replacement had anything near the same application knowledge or could run the team better. Relationships with your managers have an importance that competency at your job may not be able to replace. This again shows that a manager's success is dependent on much more than successful planning or even management execution.

Carroll and Gillen (1987) reviewed the link between traditional management functions, including planning, and success. They found that planning is correlated to both group and individual success and is one of the most important of the management functions. They report for 28 managers studied over two weeks, planning took up 19% of their work time. They also note,

Stagner (1969) found that the time 109 chief executives spent in organizational planning was related to the firm's profitability. There is some evidence that planning is important at the lowest management level as well as at the top. For example, a study of foremen at the General Electric Company (1957) revealed that foremen with higher production records spent more time in long-range planning and organizing than did foremen with poorer production records." For individual managers "Skill in planning/decision making as measured in assessment center exercises was one of the strongest predictors of managerial success." (p. 42)

This is illustrated in Table 20.1.

In a related paper, Carroll and Gillen (1984) also report that when AT&T studied the assessments of 8,000 entry-level managers, skill in planning was one of the strongest predictors of managerial success.

But is the time a manager spends planning his own work related to his success? Mankins (2004) measured the amount of time that top management teams spent on strategic planning from 187 companies worldwide with capitalization of at least $1 billion. He found that on average top management spent 37 hours per year or approximately 15% of their overall time planning. This is a significant investment of a senior manager's time.

In a paper on time management Ellwood (2005) discusses the correlation between managers' seniority and the amount of time they typically spent planning. He based the paper on 200,000 hours of real-time data gathered since 1990. See Figure 20.1.

His results show managers spend on average 8% of their time planning (Table 20.2). However, this is from a mix of senior and mid-level managers unlike Mankins' study. Ellwood also notes: "Senior managers and presidents, being higher in the organization, spend more time on planning compared with middle managers and sales managers" (p. 13). This could be consistent with Mankins' findings of 15%.

TABLE 20.1

Relationship of Managerial Skills to Unit Productivity/Efficiency

Managerial Skill	Sample 1 Manufacturing Firms (56 Units)	Sample 2 Aerospace Firms (48 Units)
Supervising skill	.46[a]	.25[a]
Planning skill	.34[a]	.43[a]
Investigating skill	.19	.20
Coordinating skill	.19	.30[a]
Evaluating skill	.10	.08
Staffing skill	.23[a]	.12

Source: After S. J. Carroll and D. J. Gillen, *Academy of Management Review* 12: 38–51, 1987.
[a] $p < .05$.

How Managers Spend Their Time

FIGURE 20.1

How managers spend their time. (After M. Ellwood, in International Association of Time Use Researchers Annual Conference, Halifax, Canada, 2005.)

TABLE 20.2

Time Spent Planning by Managers

| | **Planning Category** | | | | |
	Hours per Week	**Occasions**	**Duration in Minutes**	**Ideal Hours**	**Difference versus Ideal**
Middle Manager	4.7	10	28	5.5	−0.8
Senior Manager	9.8	18	32	7.2	+2.6
Sales Manager	4.1	14	17	4.2	−0.1
President	14.8	20	44	13.6	+1.2
All Managers	4.0	13	19	4.1	−0.1

Source: After M. Ellwood, in *International Association of Time Use Researchers Annual Conference,* Halifax, Canada, 2005.

Managers who were asked how much time they would like to spend on planning and then monitored to see how much they did spend, spent more time on an activity called long-term planning than they would like. Managers who tracked this activity spent 3.7 hours per week on long-term planning, but ideally would only like to spend 2.4 hours. At the higher level, more time is spent on long-range planning: five hours per week for presidents.

CONCLUSION

We can see from the literature that the trend is for more senior managers to spend more time planning. These data raise a question: do senior managers need to plan more because of their role or are those managers who spend more time planning most likely to become senior managers? Carroll and Gillen (1987) suggest the latter. If there is a recommendation that can be made of how much effort managers should put into planning, it is not clear from the literature. However, we can note the following conclusion:

It is not clear if planning is an important aspect of a manager's success. But we can report that the more senior the manager, the more time that is spent on planning activities.

21

New Research: Planning and Manager's Success

I've failed over and over and over again in my life and that is why I succeed.

Michael Jordan

As my final investigation I looked at whether project managers' planning habits are related to project manager success. To facilitate the analysis of the project manager's personal planning habits, two more indexes were defined:

$$\text{Personal Planning Index} = \frac{\text{number of hours spent planning own work}}{\text{Total hours worked per week}}$$

$$\text{Personal Project Planning Index} = \frac{\begin{array}{c}\text{number of hours spent on}\\\text{project planning–related tasks}\end{array}}{\text{Total hours worked per week}}$$

These indexes measure the fraction of a project manager's time that is spent planning his or her own work and time spent planning project work. They are used for measuring planning habits and personal success.

PERSONAL PLANNING HABITS

For the personal data I collected, manual review identified some outliers. The personal planning index was chosen for exclusion of outliers. Personal planning indexes less than .01 were also excluded based on the assumption that everyone must spend more than zero time planning their work. This resulted in 668 valid responses after removal of outliers.

PLANNING AND JOB FUNCTION

In Table 21.1 we examine the means for planning by job function. I attempted to find if a correlation could be found between personal planning index and job function. We can see that with a *p* of .961, there is no statistically significant relationship apparent in the data. There is clearly a relationship between project planning time and job function, however.

On average, respondents reported spending 13.5% of their week in planning. These results are between Mankins' result of approximately 15% of manager's overall time planning (Mankins, 2004) and Elwood's result of 8% (Ellwood, 2005). The 6.3 hours per week in planning is also in line with Ellwood's findings of managers spending between 4.7 to 14.8 hours per week planning (Ellwood, 2005). Interestingly, this is also relatively close to the reported mean planning phase effort for projects of approximately 15% of total effort; see Table 15.1.

This poor *p*-value may be due to the fact that the relationship is not linear. Also, the job functions as taken from the literature may not be ordinal. For example, whether a senior manager is a more senior role than a program

TABLE 21.1

Mean Planning Indexes by Job Function

Job Function	Hours per Week Planning Own Work	Personal Planning Index	Hours Spent in Project-Related Planning	Personal Project Planning Index	Responses
Project Team Member	5.96	0.14	8.69	0.19	52
Project Coordinator	6.08	0.14	12.07	0.27	60
Project Manager	6.25	0.14	14.06	0.31	284
Senior Project Manager	6.29	0.13	14.26	0.29	138
Program Manager	6.56	0.13	15.39	0.31	70
Senior Program/ Portfolio	5.86	0.12	14.05	0.28	21
Senior Manager	7.43	0.15	11.97	0.24	35
C-Level Management	6.77	0.13	13.00	0.25	13
All Groups	6.31	0.14	13.52	0.29	673
p(F)	*0.856*	*0.961*	*0.004*	*0.001*	

manager may vary between organizations and industries. If we review a graph of the means, we can see the relationship is clearly not linear thus showing why a linear regression analysis may not show a clear relationship.

The personal planning index reported by project team members and coordinators is somewhat curious in that it is higher than for most managers other than senior managers and project managers (Figure 21.1). This could be a result that reflects the findings of Mintzberg (1975) and Kotter (1999) that senior managers don't plan formally as much as react to events. Thus junior project leaders plan more and senior project managers spend more time resolving issues and removing roadblocks for their teams. Once a manager reaches senior management, planning may again become more important given the demands on the time of C-level executives (Kotter, 1999).

A nonlinear regression analysis similar to the process used for the project planning data was attempted. The results of this analysis did not appear to show a relationship, as shown in Figure 21.2. A clear curve is not apparent from this graph, and further analysis failed to find a statistically significant relationship between personal planning index and seniority. In fact, the best fit curve seems to be a straight line, showing that there is little or no relationship between personal planning time and seniority. Managers of varying seniority appear to spend an average of 13.5% of their time planning, from project coordinators to CEOs.

Of course, seniority and career development are affected by much more than just planning habits. I once consulted at a company undergoing a series

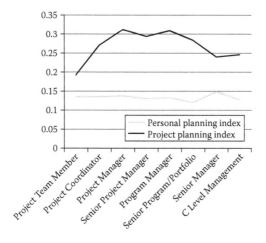

FIGURE 21.1
Planning indexes versus job function.

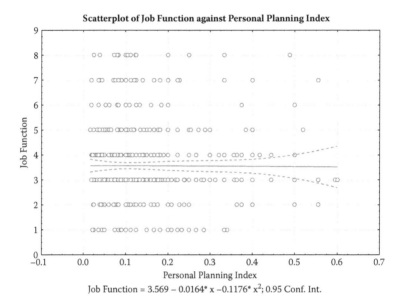

FIGURE 21.2
Scatterplot of job function against personal planning index.

of efficiency/downsizing exercises. A new large program was launched and they looked for a program manager. As a consultant, I knew I was there temporarily so I could be somewhat detached from events. Eventually, they chose an employee who was competent and a good project manager but to my eyes seemed too junior for the role. I'm sure she considered it an excellent opportunity for career advancement as did others who saw the appointment. However, the results were somewhat predictable and less than happy. The program was complex and risky and the inevitable problems arose. I wouldn't say the problems were her fault or she did a bad job; it was the nature of the program. However, when the next round of cutbacks was announced, her name was on the list. One could argue in planning her career path, she didn't plan for the risks of that appointment as well as she could have.

ADDITIONAL ANALYSIS

I took additional exploratory investigations by attempting to create personal success measures that combined some of the questions in the survey. Analysis with some of the other moderators was undertaken as well

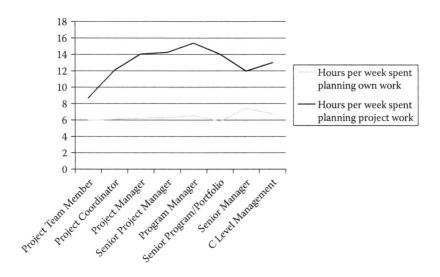

FIGURE 21.3
Hours per week versus job function.

as nonlinear analysis using log functions and cubed and higher transformations. However, all results showed both low R-values and p-values that did not indicate significance.

Next I examined the number of hours spent planning rather than the ratio (Figure 21.3). The graph of hours spent planning shows that senior managers and CEOs do spend more time planning. However, upper management also spends more time working overall. So their index value is relatively lower.

MANAGER'S PROJECT PLANNING TIME

Similarly, we can do an analysis of the personal project planning index and job function relationship shown in Figure 21.4. Even though ANOVA analysis implies there is a relationship, a binomial analysis does not reveal this. For project planning, there clearly was a significant relationship between project planning and project success whereas for personal career success, no relationship is apparent even with a large dataset. Having a large highly variable dataset alone did not ensure a relationship could be found.

I then took a further look at a regression analysis. Normal regression analysis is not suggested for analyzing ordinal dependent variables. Logit regression is indicated in those cases (Cooper and Schindler, 2008).

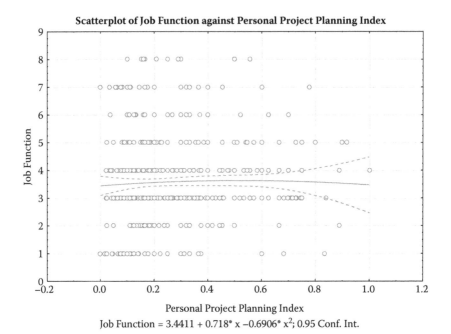

Job Function = 3.4411 + 0.718* x −0.6906* x^2; 0.95 Conf. Int.

FIGURE 21.4
Scatterplot of job function against personal project planning index.

TABLE 21.2

Logit Regression of Demographic Variables versus Job Function

Job Function—Test of All Effects	
	p
Age	0.128
Years of experience	0.000
Gender	0.227
Number of certifications	0.954
Highest education level	0.007

If we look at the other factors and their impact on seniority using a logit regression, we see the results in Table 21.2. Both years of experience and education level are correlated to job function. Those functions do have an impact on respondent's seniority as one would logically expect. This is not surprising; people with more seniority and higher education levels tend to have more senior positions.

EXPERIENCE AND PROJECT SUCCESS

Other studies such as Müller and Turner (2007) have found that project manager competencies have an impact on project success. Teague and Cooke-Davies (2007) linked project manager experience to competency. I did not study project manager competency in the survey but it may be interesting to look at experience and project success. Let us briefly examine this aspect. First we look at correlation among the project success measures and the project manager characteristics, Table 21.3.

In agreement with Teague and Cooke-Davies (2007), years of experience appears to be the item most clearly correlated with project success. Age is also significantly correlated. It is logical to assume age and years of experience would be correlated and this does turn out to be the case with a correlation of .67, $p < .05$. Field (2009) suggests that predictor correlations should be less than 0.8. The correlation between age and years of project management experience of .67 is below this threshold but close enough to question whether they should be treated separately.

The other item significantly correlated is personal planning index. However, this correlation is negative, implying that a project manager spending too much time planning her own work is detrimental to project success. However, this analysis was only conducted for the first set of projects reported by the respondent (the more successful projects).

TABLE 21.3

Correlation Analysis between Demographic Variables and Project Success Measures for "More Successful" Projects

	Correlations					
	Marked Correlations Are Significant at $p < .050$ $N = 593$					
	Job Function	**Age**	**Years of Experience**	**Number of Certifications**	**Personal Planning Index**	**Personal Project Planning Index**
Project success rating	0.09[a]	0.12[a]	0.14[a]	−0.01	−0.09[a]	−0.00
Efficiency Measure	0.06	0.10[a]	0.14[a]	0.03	−0.06	0.04
Success Measure	0.07	0.11[a]	0.11[a]	−0.01	−0.08	0.04

[a] $p < .05$.

TABLE 21.4

Correlation Analysis between Demographic Variables and Project Success Measures for All Projects

Correlations

Marked Correlations Are Significant at $p < .050$ $N = 1,112$

	Job Function	Age	Years of Experience	Number Of Certifications	Highest Education Level	Hours Worked per Week	Hours per Week Spent Planning Own Work	Personal Planning Index	Project Related Planning Time	Personal Project Planning Index
Project success rating	0.01	0.07[a]	0.09[a]	-0.06	-0.06[a]	0.03	0.03	0.00	-0.01	-0.05
Index efficiency	-0.01	0.07[a]	0.08[a]	-0.00	-0.01	0.00	0.02	0.01	-0.00	-0.01
Index success	-0.01	0.07[a]	0.07[a]	-0.05	-0.04	0.01	0.01	0.01	-0.03	-0.04

[a] $p < .05$.

Again, career success may not always correlate with project success. I recall a project manager working in a very challenging environment. His political skills were unparalleled. He managed a project that presided over the revamping of financial statements. This project resulted in errors necessitating resending 20,000 financial statements to customers, twice! Yet he managed to keep his role and his job while other project managers who had not managed such examples of failure were let go. It is hard not to agree that political skills are vitally important to a project manager's success.

Now, if the analysis is completed for all the projects reported, successful and less successful, the results are somewhat different, as shown in Table 21.4. Looking over the whole range of projects, years of experience and age are still correlated but planning habits are not. However, we need to keep in mind that these are not the average projects for the respondents but one that was successful and one that was not. A respondent could have 9 out of 10 successful projects but I have asked them to report on the one unsuccessful project and not their average projects. However, the fact that years of experience is still correlated with success shows that it is a factor. It appears to show that project managers with more experience deliver more successful "good projects" and less unsuccessful "bad projects." Again this is not surprising.

But based on the overall results of all analyses,

I cannot confirm a relationship between personal planning time and seniority or personal success.

CONCLUSION

There is little in the way of recommendations to managers that can be given regarding personal planning time. As Mintzberg (1975) and Kotter (1999) note, planning may not be the key to manager success. However, there is also no indication that managers who plan less are more successful. Planning is likely to be somewhat important to manager success per Carroll and Gillen (1987). Perhaps as in project planning, quality is more important than quantity (or time in this case). This research did not measure the quality of manager's planning efforts, and this is a potential area for future study. What can be reported is that managers of different levels report spending approximately 15% of their time in personal planning.

However, project managers could take from this research and the literature review that extra planning may not be as important to manager success as is constant monitoring and action as recommended by Mintzberg (1975) or fostering communications with senior managers per Luthans (1988) and Crawford (2000).

22

Conclusions

By failing to prepare, you are preparing to fail.

Benjamin Franklin

So there is a lot to digest. If I go back to my introduction: projects, like many things in life, require careful up-front planning. But good planning isn't the end of the story. Flexibility during execution is also important. Agile and iterative approaches show clear benefits to project success. Like canoeing, you also need to be flexible enough to handle challenges well and take advantage of opportunities.

This research did confirm the relationship between planning effort and project success. The initial findings were that a quadratic relationship exists between the percentage of effort spent planning and project success. This relationship showed statistical significance with a low p-value but also had a low R^2- value, which showed a relatively weak relationship. After completing moderator analysis, a model was derived that showed that this relationship had an R^2 of .15, which is a notable relationship for factors in the study of project management. Projects are often large complex efforts and any one factor that can account for 15% of that success is important.

There is also a clear link between iterative approaches and success. If you haven't tried iterative or agile approaches in your projects, you probably should as the more projects used agile or iterative approaches the more successful they appeared to be. After moderator effects were accounted for, an R^2 relationship of .17 was found. This has a similarly important impact on project success.

We have also learned that there can be too much planning as well as too little. Too much planning is associated with unsuccessful projects. However, it may be a symptom of a project with inherent challenges with which the team is having difficulty grappling. Executives are advised to examine projects with both too little planning or too much.

We have also found that the optimum ratio of planning to total effort was 25%. This needs to be taken with some caution as projects can vary quite a lot: no two are alike. However, it is a good rule of thumb and the best data we have so far as I have shown by the literature review.

We also found that personal planning habits are probably not the key to a manager's success. Communication with senior management, quick reaction to issues, and being active in the project seem to be more important to such success. The time that managers report they spend reflecting and planning appears not to be correlated with either greater or lesser career success. Planning their own time should probably not be the major focus of managers intent on advancing their careers.

Experienced project managers may see a hint of recognition as to what has gone well or gone badly in some of their past projects.

I hope this book has given you some information that, in the long term, you will find useful. I hope you will use the insights in this book to help improve your particular projects and perhaps even help propel your career forward.

Abbreviations

α: Alpha

ANOVA: Analysis of Variance. In the book, shown as $p(F)$ which indicates if relationships of variance of means are statistically significant.

APM: Association of Project Management (United Kingdom)

β: Beta

BOK: Body of Knowledge

CII: Construction Industry Institute

CoP: Community of Practice

CSF: Critical Success Factor

FEL: Front-End Loading

IS: Information Systems

ISO: International Organization for Standardization

IT: Information Technology

LOGIT: Logistic Regression

M: Mean

MHRA: Moderated Hierarchical Regression Analysis

OGC: Office of Government Commerce (United Kingdom)

p: The p-value is the probability of obtaining a test statistic at least as extreme as the one that was actually observed.

PDRI: Project Definition Rating Index

PM: Project Manager

PMBOK® Guide: *Guide to the Project Management Body of Knowledge* (PMI®)

PMI®: Project Management Institute

PMO: Project Management Office

PRINCE2: PRojects IN a Controlled Environment (United Kingdom)

R: Pearson's coefficient

R^2: Coefficient of determination; a measure of how well future outcomes are likely to be predicted by a model

R&D: Research and Development

RCA: Requirements Capture and Analysis

SD: Standard Deviation

SIS: Strategic Information Systems

WBS: Work Breakdown Structure

XP: Extreme Programming

Checklists

CHECKLIST 1

Artifacts to be Produced during the Planning Phase

Item	Artifact	Required (Y/N)	In Place (Y/N)	Notes
1	Project Proposal			
2	Feasibility Study			
2.1	Cost/Benefit Analysis			
2.2	Critical Success Factors			
3	Funding Request			
4	Project Charter			
5	Stakeholder Analysis			
5.1	Stakeholder Management Plan			
6	Project Communications Plan			
7	Governance Model			
7.1	Steering Committee Structure			
8	Staffing and Organization Plan			
9	Project Plan			
9.1	Estimation			
9.2	Task List or Work Breakdown Structure			
10	Risk Management Plan			
10.1	Risk Log			
11	Project Management Standards and Procedures			
11.1	Project Status Reporting Structure and Process			
12	Dependencies Analysis			
13	Change Management Plan			
13.1	Change Management Log			
14	Issue/Actions/Decisions Log			
15	Records Retention Plan			
16	Requirements Specifications			
16.1	Use Cases			
16.2	Gap Analysis			
16.3	Traceability Matrix			
17	Acquisition Plan			
18	Vendor/Supplier Management Plan			
18.1	RFP (Request for Proposals)			

Continued

CHECKLIST 1 (*Continued*)

Artifacts to be Produced during the Planning Phase

Item	Artifact	Required (Y/N)	In Place (Y/N)	Notes
19	Test Strategy			
21	Quality Assurance Plan			
22	Production Deployment Strategy			
23	Production Support Strategy			

CHECKLIST 2

Planning Phase Questionnaire

Item	Question	Applicable (Y/N)	Response
	Executive Issues		
1	How does this relate to the strategic plan?		
2	How will this affect our competitive position?		
3	Who's accountable for this project's success?		
4	Are there hard deadlines (i.e., legislated)?		
5	Is the steering committee active in project oversight?		
	Stakeholders		
6	Have the scope, objectives, costs, benefits, and impacts been communicated to all stakeholders and work groups?		
7	Have all stakeholders and work groups committed to the project?		
	Project Estimating		
8	Were multiple estimation methods employed?		
9	Will actuals be compared against estimates to analyze and correct variances?		
10	Are software metrics formally captured, analyzed, and used as a basis for other project estimates?		
11	Were project team members involved in detailed estimating and scheduling?		
12	Were stakeholders aware and supportive of the principles and practices of modern software estimation?		
	Scheduling		
13	Have adequate resources been provided by management to ensure project success?		
14	Are individual tasks of reasonable duration (8–40 hours)?		

CHECKLIST 2 (*Continued*)

Planning Phase Questionnaire

Item	Question	Applicable (Y/N)	Response
16	Are changes in deliverable commitments agreed to by all affected groups and individuals?		
17	Is an industry-recognized mechanized support tool(s) being used for project scheduling and tracking?		
18	Are internal project status meetings held at reasonable intervals?		
19	Are subproject reviews held at reasonable intervals?		
20	Have adequate procedures been put in place for project coordination and status reporting across project boundaries (i.e., other groups or departments)?		
21	What reports are sent to whom, when?		
	Risk Management		
22	Is there a process in place to monitor project risks?		
23	Will risks be reviewed regularly during the project?		
24	Have all unresolved risks been documented?		
	Have all unimplemented risk strategies been escalated to an issues log?		
	Quality Assurance		
25	Is there a quality plan covering all policies, guidelines, and procedures?		
26	Are the results of QA reviews provided to affected groups and individuals?		
27	Are adequate resources provided for the QA function?		
28	Are quality metrics defined?		
29	Is there a set of procedures to capture, analyze, and act on quality metrics?		
	Vendor Management		
30	Is there a formal set of procedures (for status reporting, contract negotiation and review, time/ invoice reconciliation, etc.) supporting vendor management?		
	Issues Management		
31	Is there a formal set of procedures supporting issues management?		

Continued

CHECKLIST 2 (*Continued*)

Planning Phase Questionnaire

Item	Question	Applicable (Y/N)	Response
	Resourcing		
34	Have arrangements been made to obtain special expertise or competence by consulting?		
35	Have the personnel with the necessary skills and competence been identified? Has agreement for their participation in the project been reached with the appropriate management?		
36	Is there a project organization chart?		
37	Has a proper project work location been established that will allow the team to work together?		
38	Does the detailed work plan match the complexity of tasks with the capabilities of personnel?		
39	Has allowance been made for vacations, holidays, training, and staff turnover?		
40	Has adequate time for orientation and training of project staff been provided?		
41	Has appropriate allowance been made for the effect of the learning curve on all personnel joining the project who do not have the required expertise?		
42	Are project leaders committed to this project full time?		
43	Are project team members committed full time?		
44	Will the production support function be adequately resourced?		
	End Users		
45	Is end user involvement adequate?		
46	Is there a service level agreement (SLA) with the appropriate departments?		
47	Are the project team members located locally to the users and have the opportunity to meet?		
48	Will users be adequately trained and are all training requirements understood?		
	Production and Operations Support		
49	Do adequate operations procedures exist?		
50	Is the production support function well defined?		
51	Are any of the following types of maintenance carried out on a planned basis:		
	a. Perfective maintenance?		
	b. Preventative maintenance?		
	c. Adaptive maintenance?		

CHECKLIST 2 (*Continued*)

Planning Phase Questionnaire

Item	Question	Applicable (Y/N)	Response
52	Are service level agreements in place between the support functions and the user departments?		
53	Is production problem resolution supported by:		
	a. Formal procedures?		
	b. Prioritization?		
	c. Accurate time and cost estimating?		
	d. Reporting?		
54	Is there an improvement program in place?		
55	Are help-desk functions well defined, efficient, and adequately resourced?		
56	What about team morale? Fun?		

Appendix A:
The Original Research

RESEARCH METHODOLOGY

The goal of the research was to understand the impact of planning on success as it is used today from a macro level. In my study a large sample of participants was desired. The responses were to be analyzed using statistical methods. Moderating variables were also to be analyzed. Both goals required a large data sample and maximizing the number of participants who could provide quality data.

Figure A.1 is a diagram showing the relationship between the planning effort and the success metrics. The moderating variables affect that relationship. The secondary construct of the research had the structure shown in Figure A.2.

Martin, Pearson, and Furumo (2005) report that project size, experience with a technology, project complexity, use of outside vendors, and technological complexity all have impacts on project success. These items were all considered as potential moderating factors and were therefore included in the data collection. Many additional moderators to project success were reported by Shenhar et al. (2001), Premkumar and King (1991), Müller and Turner (2007), and Pinto and Prescott (1990), and were included in the research. The moderating variables shown in Table A.1 were analyzed.

RESEARCH CONSTRUCTS

In measuring the effort expended in planning only, I departed from much of the literature in this area which had a greater focus on quality of planning. Quality measures are somewhat subjective and there is no

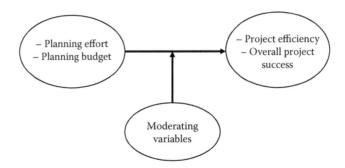

FIGURE A.1
Main research constructs.

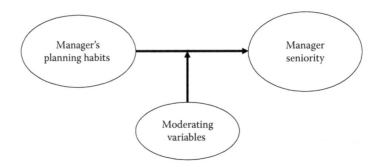

FIGURE A.2
Secondary research constructs.

guarantee that perceptions of quality in planning outputs are comparable among industries. To truly compare the quality of planning would require a breakdown of the creation of planning deliverables as in Dvir et al. (2003). However, from Table A.2 you can see that not all planning deliverable breakdowns will apply to all industries or projects. ILS (integrated logistics support), for example, is likely to be unfamiliar to many readers and other readers may find many items missing from this list.

There are risk areas. For example, a 10% planning effort (as a fraction of total project effort) by one team may not be the equal of a 10% effort by another team. Teams have different levels of training, experience, and effectiveness. This effect could be reduced by the likelihood that a team that is less efficient in planning may also be less efficient overall. Therefore, the percentage would be less affected, and the ratio of planning effort to the project total was still of interest. That is one of the reasons that I chose to focus on analyzing the ratio between planning and efforts

TABLE A.1

Moderating Variables

	Moderating Variable	Rationale	Reference
1	Project team size	Larger teams could potentially require more planning. Is there an impact on project success?	Crawford, Hobbs, and Turner (2004)
2	Project complexity	Complex projects could require more planning than less complex projects. What is the impact on project success?	Dulewicz and Higgs (2005), Martin et al. (2005)
3	Project length	Do shorter projects require more planning or do longer, bigger projects require more planning? What is the impact on project success?	Cooke-Davies (2002), Martin et al. (2005)
4	The detail level of the WBS used in the project	The WBS is part of the planning process but does it have a direct effect on planning effort or project success?	Tausworthe (1980)
5	The applicability/quality of the goals/ vision statement for the project	The goals and vision statement is defined as part of the planning process but do they have a direct effect on planning effort or project success?	Pinto and Prescott (1990), Christenson and Walker (2008)
6	Novelty to the organization: how new is this type of project to the organization?	Could novel projects require more planning to be successful?	Shenhar et al. (2001)
7	Internal versus vendor based	Are vendor-based projects more risky and therefore in need of greater planning, or does the vendor take on some of the planning role?	Crawford et al. (2004), Martin et al. (2005)
8	Industry	Is there a clear difference in the amount of planning required in different industries, or is there some consistency across industries?	Shenhar et al. (2001), Premkumar and King (1991)

Continued

TABLE A.1 (*Continued*)
Moderating Variables

	Moderating Variable	Rationale	Reference
9	Geographic location of project	Is there a difference in how much planning is done in different parts of the world and is this reflected in the perceived success of those projects?	Turner (2000)
10	Local versus remote team	Do geographically dispersed teams require more planning? What is the impact on project success?	Crawford (2001)
11	Level of use of technology	Is it important to plan more in high-technology projects?	Shenhar et al. (2001)
12	New product versus maintenance	Do projects such as maintenance which may be considered lower risk, require less planning? What is the impact on project success?	Shenhar et al. (2001)
13	Experience level of team	Do more experienced teams typically do more planning, or do they require less planning? What is the impact on project success?	Scott-Young and Samson (2008)
14	Degree of stakeholder engagement	Is the level stakeholder engagement and its quality important to planning and to project success?	Premkumar and King (1991)
15	Methodology type: How much of the project was done using agile or iterative techniques?	Methodology type is expected to have some impact on the amount of planning effort. Is total planning effort in agile projects less than in traditional methodologies and by how much?	Ceschi et al. (2005)
16	Project size	Do larger projects require a higher percentage of the planning effort compared to total effort or is the ratio relatively constant?	Crawford et al. (2004)
17	Project budget	Do projects with higher budgets require a higher percentage of the planning effort compared to total effort or is the ratio relatively constant?	Crawford et al. (2004)

TABLE A.2

Measures for the "Implementation of Project
Management Processes and Procedures" Items

1.	Systems engineering
2.	Engineering design
3.	Risk management
4.	Resource and schedule planning
5.	Financial management
6.	Contract management
7.	Procurement management
8.	ILS management
9.	Quality and reliability assurance
10.	Test and inspection management
11.	End user relationship management
12.	Configuration management
13.	Change management
14.	Team management
15.	Meeting and decision-making management
16.	Reporting and communications
17.	Transfer to production

Source: After D. Dvir, T. Raz, and A. Shenhar, *International Journal of Project Management* 21: 89–95, 2003.

compared to total project effort. From the literature, we can see that the impact of ratio of the planning effort to the total of project success has not been studied previously.

DATA GATHERING: UNIT OF ANALYSIS

The unit of analysis for this research is a completed project. As defined by Turner et al. (2010), PMI® (2013), a project is a temporary organization to which resources are assigned to deliver benefits. Projects were studied that have information available on the effort and budget spent on the planning phase. Projects needed to have been completed so that success could be fully judged.

Both effort and budget were to be analyzed, but the main focus was on analyzing planning effort. It was assumed that effort data would be more readily available and be more accurate as some projects have substantial equipment and materials costs embedded in budgets (Wideman, 2000).

In addition, respondents were asked about demographics details and planning habits. Therefore the secondary unit of analysis was the individuals who responded to the survey.

SAMPLING FRAME

The type of project studied, location, or industry was not constrained. The questionnaire was open to project managers in a variety of industries and managing a variety of project types. Differences in the projects were captured and were analyzed as moderating variables. The goal was to see if there was any relationship that could be defined about planning in projects in general as well as to understand how characteristics of different projects affect the need for planning. For that reason a large number of participants from a wide variety of different industries and countries were sought.

Data were collected through web-based surveys of project management practitioners. The following describes the participants:

- Project Management Institute (PMI): There were several avenues used to reach PMI members.
  - PMI survey links site.
  - Community of Practice (CoP) message boards.
  - A PMI local chapter also agreed to include a link to the survey in its weekly e-mail newsletter.
- LinkedIn groups with a project management focus had posts added to their discussion boards requesting participants.
- Project managers in the researcher's personal network were contacted and asked to participate. This resulted in a small number of additional participants.

Sample Sizing

Identifying the resulting sample size was not possible. Although the membership numbers for the LinkedIn groups are available (typically in the 1,000s) and membership numbers in the PMI Communities of Practice

(CoPs) are also available (membership up to the 10,000s), memberships in each of these groups are not mutually exclusive. Project managers may be members of many LinkedIn groups and PMI CoPs. As well, how many members read postings on the discussion boards is not known. Finally, the actual number of people who saw the posting for the survey or received an invitation by e-mail cannot be quantitatively determined.

The overall goal of the questionnaire process was to maximize the amount of project data available to study. The literature review revealed differences in planning requirements and success rates in different industries (Collyer et al., 2010) and even in different parts of the world (Crawford, 2000). To be able to distinguish planning effects from effects related to project characteristics, I felt that a large sample size would be required.

The full survey along with further information on how it was structured can be found in Appendix C.

The goal of the data collection was to get as large a dataset as practical. Based on the literature reviewed, it was expected that projects would vary greatly in planning characteristics and that a large number of projects might be necessary to detect trends. According to Denscombe (2007), one should only expect a small percentage response from every contact made.

Data collection commenced in fall 2011 and was completed in early 2012 for a collection period of approximately 12 weeks. A total of 865 people started the survey with 859 completing at least the first portion of it, which requested information on one more successful project. Table A.3 is a summary of the source of the participants.

TABLE A.3

Sources of Survey Participants

Source of Participants	Number Who Started Survey	Number Who Completed Survey
PMI Survey Links Site	18	18
PMI Communities of Practice	96	96
PMI Local Group	0	0
PMI Information Systems Community of Practice E-Mail	542	539
LinkedIn Groups	197	194
Personal Network	12	12
Total	*865*	*859*

Each participant was asked to provide data on two projects, one more successful and another less successful. However, not all participants entered data for two projects; therefore, the total number of projects available for study was 1,539.

Fellows and Liu (2003) note that collecting data is becoming progressively more difficult given the increasing number of research projects being conducted. The respondents are being bombarded with many requests for data and are therefore becoming unwilling to spend a lot of time on them and ultimately decline to participate in academic surveys. Olomolaiye (2007) highlights that lack of understanding of the value of the research area can also lead to poor participation. Given these challenges, I was satisfied with the relatively large number of responses received to the survey.

Figure A.3 gives a breakdown of the country of the respondents. For purposes of this study, this refers to the country where the respondent currently resides. Some chose not to respond to this question as all demographic questions were optional.

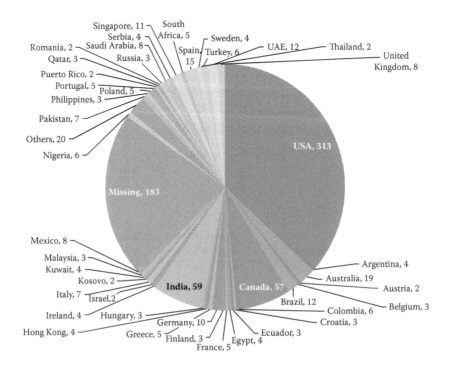

FIGURE A.3
Reported country of origin of respondents.

We can see that a large percentage of respondents were from the United States (36%), but there is still a global level of participation in the survey with India, Canada, Australia, Spain, Brazil, UAE, Singapore, and Germany having 10 or more respondents. The group labeled as others, consisted of 20 countries and contains one response each from: Bulgaria, China, Costa Rica, Ethiopia, Iran, Jamaica, Japan, Jordan, Lebanon, Mongolia, Norway, Sri Lanka, Sudan, Switzerland, Syria, Taiwan, Ukraine, Venezuela, Vietnam, and Yemen. Overall, there was a global representation with respondents from more than 60 countries.

Validity

As noted in Cooper and Schindler (2008), it is important to design survey questions such that validity is adequate. This design often involves the judgment of the researcher. In this survey, the key independent variable is planning effort; rather than ask the percentage of time spent planning on a project, respondents were asked to provide effort and budget values for the planning phase. The intent was to require additional investigation from the respondent by ensuring they looked at actuals rather than rely only on memory.

Other questions in the survey relied on questions used previously in the literature to minimize any concerns on validity. Where questions were not found in the literature, new questions were designed in keeping with what was found in the literature review; see Table C.1 in Appendix C.

Participant Error Issues

Surveys can suffer from participant errors (Cooper and Schindler, 2008). These can include nonresponse errors where the participants fail to respond to particular questions, and response errors where the participant does not give an accurate response or gives an incomplete response. These errors were of particular concern relating to the quantitative data on the planning phase and total project effort and budget. This information may not be readily available for all project managers to access. In manually reviewing the data, it was clear that some participants did not have access to these data as zero was given as a value. In other cases the project totals were entered as the same as the planning phase totals or the planning phase was entered as an unrealistically high percentage of the total. Steps were taken during analysis to remove those entries.

Another source of error noted by Cooper and Schindler (2008) is when participants interpret the questions differently than what was intended by the researcher. This was also found to be the case in the two parts of the survey. Some participants entered projects in the "more successful" section of the survey that they ranked as failures, and projects in the "less successful" section they ranked as highly successful (45 of 859 successful projects and 97 of 722 less successful projects, respectively). However, these cases were left in as the key measure for success was in the eight questions asked relating to project success and not which section the information was entered. Therefore for the majority of the analysis, data for all the projects were combined into one set. Initially, this resulted in a dataset covering 1,579 projects, although after the removal of logically invalid data and outliers the usable total was 1,386 projects. Misinterpretation of questions may have also been an issue with some of the data removed as outliers. However, after the removal of these data, there remained a substantial amount of data that was adequate to complete the research.

Two other sets of data suffered from lower rates of user response: agile planning data and budget data. This was to be expected; some projects do not use agile and therefore have no data to report. Similarly, some projects do not track budgets closely. This is often the case in internal projects, where effort is tracked but because the customers are internal, budget may not be tracked. In other cases, budget numbers were not managed by the project manager but were confidential at the senior account level. These cases were reported in the comments sections of the survey by some respondents. Therefore, for analysis of agile effort or budget effort, only the subset of the valid 1,386 cases which also had valid agile or budget data were used (938 and 1,037, respectively).

Unanswered questions were generally not an issue with this research. The majority of project-related questions were mandatory in the questionnaire. The only questions that were optional were budget amounts, agile amounts, and any demographic data. It was understood that both agile and budget data would not be available for all projects. Otherwise, blank questions were not an issue for the analyses.

Biased recall can be a concern in any study of this nature. Fischoff (1991) warns against misinterpretation of surveys when participants' values are not fully articulated, and of course this cannot be done using a survey. There is a question as to whether participants will enter valid data or data they believe the questioner wants or data that are expedient to complete the survey (Fischoff, 1991). However, the large amount of data available through creating a widely distributed survey cannot be achieved using techniques such as interviews. Therefore I was aware of these possibilities and mitigated them by testing validity and reliability.

Appendix B: Details of the Statistical Analysis

SUCCESS MEASURE ANALYSIS

For success, three items measured project efficiency, and four measured overall project success. All measures of success showed a high Cronbach alpha score which shows they are all correlated. This fits with the results of our review of the success literature and the view that all success measures are to some extent correlated (Dvir et al., 2003; Prabhakar, 2008; Kloppenborg, Manolis, and Tesch, 2009; Zwikael and Globerson, 2006). Table B.1 shows the results of this analysis.

Based on this result there is no strong reason to exclude any of the items. The average is .905 and alpha will only improve marginally by deleting project budget goals.

This also indicates that all of the factors above are interrelated to some extent. However, we can note that the only factors close to the threshold for removal are budget goals, scope, and time goals. These are key components of efficiency. Scope is the lowest of this category which is in keeping with Shenhar et al. (1997) who stated that scope was the most important of the triple constraints for overall success.

The results of the Cronbach's alpha analysis supported the initial assumptions that the elements identified for measuring success (Müller and Turner, 2007; Shenhar et al., 2001; Dvir et al., 2003; Zwikael and Globerson, 2006) were valid measures of success for this survey and accurately measured the judgments of respondents. Each variable achieved a high alpha score greater than $\alpha = .85$. In practical terms, this meant there was a high degree of confidence in the reliability and the data collected through the survey, and they are accurate and meaningful for the purposes of this research.

TABLE B.1

Cronbach Alpha Analysis of Success Measures

<table>
<tr><td colspan="2" align="center">Summary for Scale: Valid N:1,378
Cronbach Alpha: .905 Standardized Alpha: .922</td></tr>
<tr><td></td><td align="right">**Alpha if Deleted**</td></tr>
<tr><td>Project success rating</td><td align="right">0.885</td></tr>
<tr><td>Project sponsors and stakeholders
  success rating</td><td align="right">0.884</td></tr>
<tr><td>Project budget goals</td><td align="right">0.912</td></tr>
<tr><td>Project time goals</td><td align="right">0.903</td></tr>
<tr><td>Scope and requirements goals</td><td align="right">0.900</td></tr>
<tr><td>Project team's satisfaction</td><td align="right">0.888</td></tr>
<tr><td>Client's satisfaction</td><td align="right">0.884</td></tr>
<tr><td>End users' satisfaction</td><td align="right">0.889</td></tr>
</table>

Factor Analysis

In order to identify patterns of interrelationships among the variables thought to be potential moderators, a factor analysis was completed. Factor analysis is a technique for investigating whether a number of variables of interest are related to a smaller number of unobserved uncorrelated variables (Cooper and Schindler, 2008). Factor analysis assumes the following:

- Multicollinearity
- Homogeneity of sample

Given the size and nature of the sample, these items were not a concern. A normalized varimax rotation was selected to achieve the highest loadings and best model fit. A scree plot was created to confirm three factors were appropriate, Figure B.10, at end of Appendix B.

Table B.2 is a summary of this analysis. Three factors were found to be significant. Factor 2 consists of three of four of the items identified as planning factors in the hypotheses plus team experience identified within the factor analysis. Next, Cronbach's alpha analysis was conducted on the components of the three identified factors.

The Cronbach alpha coefficient is a number that ranges from 0 to 1. A value of 1 indicates that the measure has perfect reliability. A value of 0 indicates that the measure is not reliable and variations are due to random error. Ideally the alpha value should approach 1. In general, an alpha value

TABLE B.2

Factor Loadings for Moderator Variables

	Factor Loadings (Varimax Normalized) Extraction: Principal Components		
	Factor 1	**Factor 2**	**Factor 3**
Project team size	0.764*	0.183	−0.022
Complexity of the project	0.605*	0.051	−0.368
Project length	0.709*	0.041	−0.008
Detail level of the WBS	0.006	−0.741*	0.068
Quality of goals/vision	0.104	−0.733*	0.113
Novelty to the organization	−0.104	0.051	0.695*
Internal versus vendor-based	−0.502*	0.086	0.089
Level of use of technology	0.232	0.128	−0.551*
New product versus maintenance	0.005	0.015	0.708*
Experience level of team	−0.116	−0.534*	−0.189
Degree of stakeholder engagement	−0.104	−0.646*	0.040
Methodology type	−0.023	−0.264	0.242
Explained variance	1.804	1.922	1.546

*Marked loadings are > .500

TABLE B.3

Cronbach's Alpha Analysis for Factor 1

	Summary for Scale: Mean = 8.739 Std.Dv. = 2.977 Valid N:1,386 Cronbach Alpha: .490 Standardized Alpha: .585 Average Interitem Corr.: .263				
	Mean if Deleted	**Var. if Deleted**	**St.Dev. if Deleted**	**Itm-Totl - Correl.**	**Alpha if Deleted**
Project team size	6.183	4.558	2.135	0.399	0.290
Complexity of the project	6.381	7.307	2.703	0.353	0.427
Project length	7.180	7.269	2.696	0.353	0.424
Vendor versus internal	6.473	4.180	2.045	0.253	0.532

of 0.9 is required for practical decision-making situations and a value of 0.7 is considered to be sufficient for research purposes (Nunnally, 1978). In Table B.3 we can see that for factor 1, Cronbach's alpha is relatively low at .489. Internal versus vendor was recoded to remove the negative correlation identified in the factor analysis. If we delete vendor versus internal, the alpha would increase to .53. Nunnally (1978) stated that .70 is an adequate reliability coefficient, but lower thresholds have sometimes been used.

However, that is still substantially above .53, therefore analysis for this factor will not continue. See Table B.4.

In this case, we have an overall Cronbach's alpha of .61 with no benefit to be gained from removing any of the factors. This is somewhat below the typical threshold. However, inasmuch as this is not central to the hypothesis of this book but is of interest, it is at an acceptable value for this purpose (Field, 2009). We continue with some further analysis of this factor, the planning factor, later in this appendix.

In Table B.5, for the third factor, the alpha is quite low and cannot be adequately improved, so it is not further analyzed.

TABLE B.4

Cronbach's Alpha Analysis for Factor 2

Summary for Scale: Mean = 7.987 Std.Dv. = 1.994 Valid *N*:1,384 Cronbach Alpha: .608 Standardized Alpha: .604 Average Interitem Corr.: .278					
	Mean if Deleted	**Var. if Deleted**	**St.Dev. if Deleted**	**Itm-Totl - Correl.**	**Squared - Multp. *R***	**Alpha if Deleted**
Detail level of the WBS	5.668	2.316	1.522	0.459	0.239	0.480
Quality of goals/vision	5.871	2.383	1.544	0.442	0.228	0.494
Experience level of team	6.236	2.935	1.713	0.280	0.080	0.609
Degree of stakeholder engagement	5.668	2.316	1.522	0.459	0.239	0.480

TABLE B.5

Cronbach's Alpha Analysis for Factor 3

Summary for Scale: Mean = 6.973 Std.Dv. = 1.294 Valid *N*:1,386 Cronbach Alpha: .439 Standardized Alpha: .449 Average Interitem Corr.: .214					
	Mean if Deleted	**Var. if Deleted**	**St.Dev. if Deleted**	**Itm-Totl - Correl.**	**Alpha if Deleted**
Novelty to the organization	4.083	1.193	1.092	0.275	0.345
New product versus maintenance	4.676	1.646	1.283	0.297	0.307
Technology level	3.571	1.587	1.260	0.246	0.379

DEMOGRAPHICS OF RESPONDENTS

Respondents shown in Table B.6 were also asked to provide their years of experience in project management. Tables B.7 through B.10 summarize their responses.

TABLE B.6

Demographics of Participants

Category	Question	Participants	Percentage
Job Function	Project Team Member	58	6.75
	Project Coordinator	66	7.68
	Project Manager	304	35.39
	Senior Project Manager	141	16.41
	Program Manager	72	8.38
	Senior Program/Portfolio	22	2.56
	Senior Manager	36	4.19
	C-Level Management	14	1.63
	Not answered	146	17.00
	Cumulative	859	100
Age	<30	64	7.45
	31–40	262	30.50
	41–50	223	25.96
	51–60	133	15.48
	61+	29	3.37
	Not answered	148	17.22
	Cumulative	859	100
Gender	M	553	64.37
	F	154	17.92
	Not answered	152	17.69
	Cumulative	859	100
Project management certification[a]	PMP	618	71.94
	Prince2	20	2.33
	IPMA	3	0.35
	APMC	0	0.00
	AAPM	5	0.58
	AIPM	2	0.23
	PgMP	6	0.70
Education	High school	43	5.00
	Bachelor's degree	263	30.61
	Master's degree	376	43.77
	Doctorate	28	3.25
	Not answered	149	17.34
	Cumulative	859	100

[a] Note that certification percentages do not total to 100% as some have no certification and others have several.

TABLE B.7

Years of Respondent Experience in
Project Management

Years of Experience	Count	Percent
0	3	0.35
1	9	1.05
2	22	2.56
3	27	3.14
4	40	4.66
5	69	8.03
6	47	5.47
7	49	5.70
8	41	4.77
9	15	1.75
10	96	11.18
11	15	1.75
12	36	4.19
13	8	0.93
14	12	1.40
15	75	8.73
16	8	0.93
17	6	0.70
18	9	1.05
19	1	0.12
20	44	5.12
21	3	0.35
22	3	0.35
23	3	0.35
25	36	4.19
28	2	0.23
30	14	1.63
32	1	0.12
33	1	0.12
35	9	1.05
36	2	0.23
40	1	0.12
45	1	0.12
Missing	148	17.11
Total	*859*	*100.00*

TABLE B.8

Logit Regression Analysis of Methodology Type versus Project Success Rating

	Project Success Rating—Test of All Effects Ordinal Multinomial Link Function: LOGIT		
	Degree of Freedom	**Wald—Stat.**	**p**
Intercept	4	1422.172	0.000000
Methodology type	5	43.398	0.000000

TABLE B.9

Planning Effort Index Analysis of Normality

	Descriptive Statistics						
	Valid **N**	**Mean**	**Minimum**	**Maximum**	**Std.** **Dev.**	**Skewness**	**Kurtosis**
Planning effort index	1386	0.153	0.010	0.600	0.116	1.153	1.179

TABLE B.10

Planning Budget Index Analysis of Normality

	Descriptive Statistics						
	Valid **N**	**Mean**	**Minimum**	**Maximum**	**Std.** **Dev.**	**Skewness**	**Kurtosis**
Planning budget index	1109	0.128	0.010	0.600	0.111	1.504	2.370

Normal probability plots, *p–p* plots, and homoscedasticity plots are shown in Figures B.1–B.9. A scree plot is shown in Figure B.10.

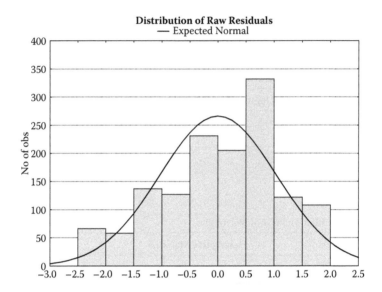

FIGURE B.1
Distribution of residuals for planning effort index versus overall success measure.

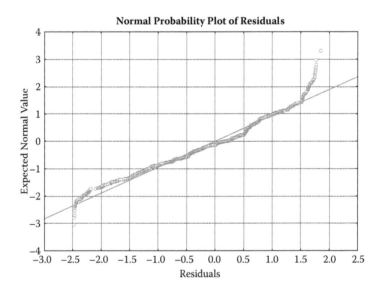

FIGURE B.2
p–p Graph of residuals for planning effort index versus overall success measure.

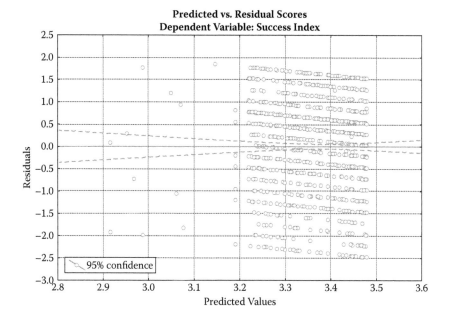

FIGURE B.3
Homoscedasticity plot of residuals for planning effort index versus overall success measure.

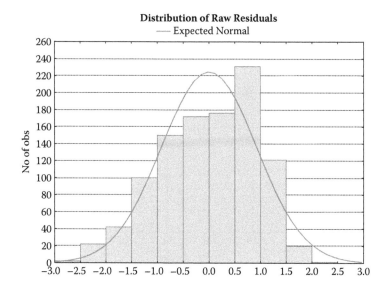

FIGURE B.4
Distribution of residuals for planning budget index versus overall success measure.

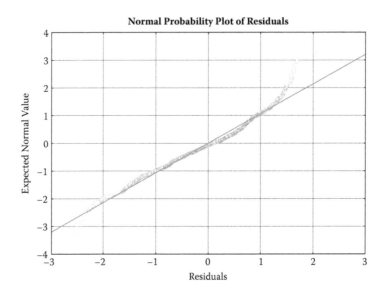

FIGURE B.5
p–p Graph of residuals for planning budget index versus overall success measure.

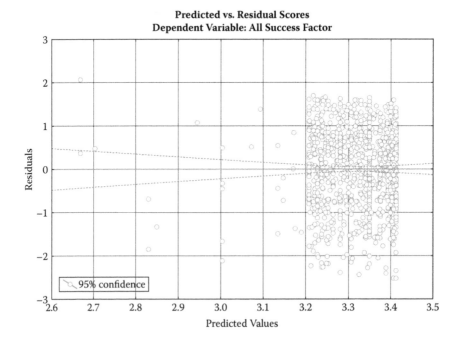

FIGURE B.6
Homoscedasticity plot of residuals for planning budget index versus overall success measure.

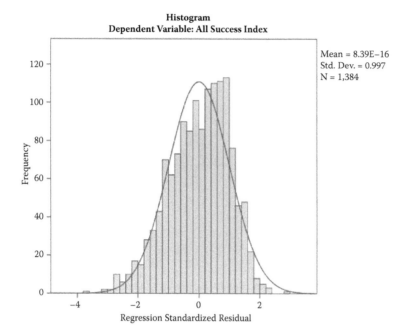

FIGURE B.7
Distribution of residuals for final planning effort model versus overall success measure.

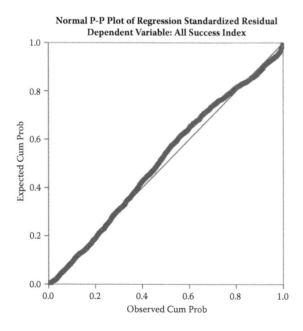

FIGURE B.8
p–p Graph of residuals for planning effort model versus overall success measure.

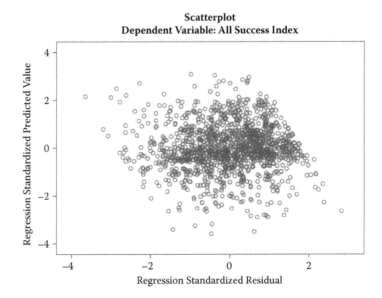

FIGURE B.9

Homoscedasticity plot of residuals for planning effort model versus overall success measure.

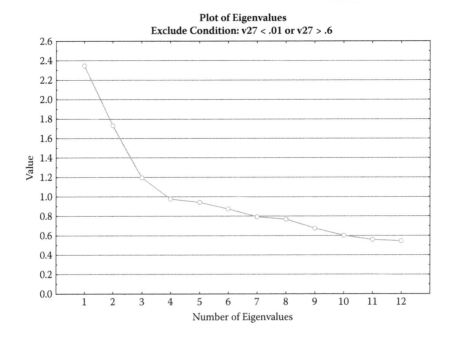

FIGURE B.10

Scree plot of moderator factor analysis.

Appendix C: Details of the Survey

WEB-BASED QUESTIONNAIRE

The following information was captured in the survey. Questions were used from relevant, previously tested questionnaires as much as possible, and the references are included in Table C.1. Where appropriate questions could not be found, they were generated by the researcher in accordance with the findings from the literature review.

Monosource bias and other response biases can occur in self-rated performance measures as discussed by Podsakoff, MacKenzie, Lee, and Podsakoff (2003) and Conway and Lance (2010). For that reason, anonymity was allowed in the survey and company names were not captured. Questions also used varied scales as recommended by Podsakoff et al. (2003). In addition, factor analysis and Cronbach's alpha analysis were completed where appropriate. To avoid social desirability issues related to project success, respondents were asked to provide data for both a more successful and less successful project. Finally, the use of PMI® groups, LinkedIn, and personal contacts ensured there were no convenience sample issues. Therefore monosource bias was assumed not to be an issue for this research.

Survey questions, in general, used a 5- or 7-point Likert-like numeric scale (Cooper and Schindler, 2008). Pure Likert scales were not used as there were several questions where numerical responses were appropriate such as team size ranges. Where appropriate, participants were asked to provide actual numerical data such as project effort and budgets. The varying scale was partially due to following the scales from existing literature, using 7-point scales to allow optimum ordinal value for numeric ranges and 5-point scales for subjective ratings. Because a variety of scales was used, this ensured that item context effects per Podsakoff et al. (2003) were not a concern even though this was a relatively long survey.

TABLE C.1

Questionnaire

The following questions will be asked twice:

Category	ID	Question	Response Ranges	Reference
Quantitative Data	16	Project effort What was the total project effort (in person-days)?	Numerical: person-days	Crawford et al. (2004)
	17	Planning Phase effort What was the total effort expended on the planning phase (person-days)? Planning is defined as everything before execution.	Numerical: person-days	Crawford et al. (2004)
	18	Total effort spent on planning over life of project. For Agileagile/Iterative projects, what is the total effort expended in planning over the course of the project? This should include all planning down to weekly planning sessions.	Numerical: person-days	Crawford et al. (2004)
	19	Project Budget What was the overall project budget? (USD: please do the approximate conversion to US dollars).	Numerical Approximate in US dollars	Crawford et al. (2004)
	20	Planning Phase Budget What was the total budget for the planning phase (US dollars)? Planning is defined as everything before execution.	Numerical Approximate in US dollars	Crawford et al. (2004)
	21	Total budget spent on planning over life of project For Agileagile/Iterative projects, what is the total cost (USD) expended in planning over the course of the project? This should include all planning down to weekly planning sessions.	Numerical Approximate in US dollars	Crawford et al. (2004)

Success Rating	22	Project success rating: overall How do you rate the overall success of the project?	5-point scale: failure not fully successful mixed successful very successful	Shenhar et al. (2001)
	23	Project success rating: sponsor feedback How did the project sponsors and stakeholders rate the success of the project?	5-point scale: failure not fully successful mixed successful very successful	Shenhar et al. (2001), Müller and Turner (2007)
	24	Success: Meeting budget goals How successful was the project in meeting project budget goals?	7-point scale: >60% over budget 45–59% over budget 30–44% over budget 15–29% over budget 1–14% over budget on budget under budget	Shenhar et al. (2001), Dvir et al. (2003), Zwikael and Globerson (2006)

Continued

TABLE C.1 (*Continued*)

Questionnaire

Category	ID	Question	Response Ranges	Reference
The following questions will be asked twice:				
	25	Success: Meeting timeline goals	7-point scale:	Dvir et al. (2003)
		How successful was the project in meeting project time goals?	>60% over time	Zwikael and
			45–59% over time	Globerson (2006)
			30–44% over time	
			15–29% over time	
			1–14% over time	
			on time	
			ahead of schedule	
	26	Success: Meeting scope/requirements goals	7-point scale:	Dvir et al. (2003)
		How successful was the project in meeting scope and requirements goals?	>60% requirements missed	
			45–59% requirements missed	
			30–44% requirements missed	
			15–29% requirements missed	
			1–14% requirements missed	
			requirements met	
			requirements exceeded	
	27	How do you rate the project team's satisfaction with the project?	5-point scale:	Müller and Turner
			failure	(2007)
			not fully successful	
			mixed	
			successful	
			very successful	

28	How do you rate the client's satisfaction with the project's results?	5-point scale: failure not fully successful mixed successful very successful	Müller and Turner (2007)
29	How do you rate the end users' satisfaction with the project's results?	5-point scale: failure not fully successful mixed successful very successful	Müller and Turner (2007)
Moderating variables			
1	Project team size How large was the project team (full-time staff equivalent)?	7-point scale: 1–5 6–15 16–30 31–50 51–100 101–500 501+	Crawford et al. (2004)
2	Project complexity Rate the complexity of the project.	3-point scale: Low Medium High	Dulewicz and Higgs (2005)

Continued

TABLE C.1 (*Continued*)

Questionnaire

Category	ID	Question	Response Ranges	Reference
The following questions will be asked twice:				
	3	Project length What was the project length (full life cycle)?	3-point scale: • <1 year • 1 to 3 years • >3 years	Cooke-Davies (2000)
	4	WBS Rate the detail level of the WBS (work breakdown structure) used for the project.	4-point scale: Excellent Good Poor Very poor/Not used	
	5	Goals/vision statement Rate the applicability/quality of the vision statement or project goal definition for the project.	4-point scale: Excellent Good Poor Very poor/Not used	Pinto and Prescott (1990), Crawford et al. (2004)
	6	Novelty to the organization How new is this type of project to the organization?	4-point scale: Very Somewhat Mostly not Not at all	Shenhar et al. (2001)

| 7 | Internal versus vendor based: What percentage of project was completed by vendors? | 6-point scale:
80–100%
60–79%
40–59%
20–39%
1–19%
0% | Crawford et al. (2004) |
| 8 | Industry
What industry was the project in? (Choose the best fit.) | Education
Retail
High technology
Financial services
Manufacturing
Utilities
Healthcare
Government
Professional services
Telecommunications
Construction
Other | Shenhar et al. (2001),
Premkumar and King
(1991),
Catersels et al. (2010) |

Continued

TABLE C.1 (Continued)

Questionnaire

The following questions will be asked twice:

Category	ID	Question	Response Ranges	Reference
	9	Geographic location of project What was the geographic location of project?	Europe Indian subcontinent Russia and FSU Far East North America Australasia Latin America Pacific Middle East Arctic and Antarctic Africa sub-Sahara	Turner (2000)
	10	Local versus remote team Where were the team members located? Choose the option that best fits the majority of team members.	3-point scale: • One city or region • National • International	Crawford (2001)
	11	Level of use of technology Low tech indicates none or very mature technology where super-high tech indicates the use or development of completely new technology.	4-point scale: Low tech Medium tech High tech Super-high tech	Shenhar et al. (2001)

#		Scale	Source
12	New product versus maintenance Does this project involve developing a new product, installation, or system or is it related to maintenance of what already exists?	3-point scale: New product development Product modifications Maintenance only	Shenhar et al. (2001)
13	Experience level of team How experienced was the project team?	3-point scale: High Medium Low	
14	Degree of stakeholder engagement How engaged were the key stakeholders for the project?	4-point scale: Very Somewhat Mostly not Not at all	Premkumar and King (1991)
15	Methodology type: How much of the project was done using agile or iterative techniques? (100 = fully agile, 0 = fully waterfall, 50 = an equal mix of agile and waterfall techniques)	6-point scale: 80–100% 60–79% 40–59% 20–39% 1–19% 0%	

Continued

TABLE C.1 (*Continued*)

Questionnaire

Category	ID	Question	Response Ranges	Reference
Demographic data is collected for the respondents to the survey. The following information will be requested:				

Respondent Demographics	59	Job function
		What is your job function?
			8-point scale:
			Project Team Member
			Project Coordinator
			Project Manager
			Senior Project
			  Manager
			Program Manager
			Senior Program/
			  Portfolio
			Senior Manager
			C-Level Management
				Müller and Turner (2007),
				Crawford et al. (2004)

	60	Age What is your age?	5-point scale: <30 31–40 41–50 51–60 61+	Müller and Turner (2007)
	61	Years of experience in project management How many years of experience do you have in project management?	Numerical	Müller and Turner (2007)
	62	Gender What is your gender?	M F	Müller and Turner (2007)

63	Project management certification What project management certifications do you possess? Check all that apply.	PMP Prince2 IPMA APMC AAPM AIPM PgMP	Müller and Turner (2007)
64	What is your highest education level?	4-point scale: High school Bachelor's degree Master's degree Doctorate	Müller and Turner (2007)
PM Planning			
65	Average hours per week worked How many hours per week do you work on average?	Numerical	
66	Hours per week spent planning own work Estimate how many hours per week you spend planning your own work. This does not include planning projects or planning for an enterprise but planning for your own personal work time.	Numerical	
67	Hours per month spent planning How much time do you spend in planning that is project-related? This includes planning projects, updating project plans, and detailed task planning.	Numerical	
68	Contact information Please enter your contact information if you wish to be entered in the draw or to receive the results from this study.	Name: State/Province: ZIP/Postal Code: Country: E-mail Address: Phone Number:	Müller and Turner (2007)

The survey is divided into three sections. In the first two sections, users were asked to fill in two project examples, one they rated a more success-ful project and one they rated as less successful. Finally, demographic data as well as data on the participant's perceptions of his or her own planning styles were collected in the third part.

Appendix D: Final Model Sensitivity Analysis

INTERPRETATION OF COEFFICIENTS AND SENSITIVITY ANALYSIS

In addition, some sensitivity analysis was undertaken on this model. Planning index versus success measure for the final model sensitivity analysis for all three moderators was conducted, as shown in Figure D.1.

The model seems to break down somewhat for the extreme low values for team experience and detail of WBSs. These projects are a minority, accounting for 11 and 8% of projects, respectively. Because these values are related to problem projects (i.e., projects with an inexperienced team and with a very poor/nonexistent WBS), this was deemed acceptable because in those cases project failure is likely. As well, in this case the model tends to portray that planning efforts up to 60% are beneficial, which can be argued logically: these projects would benefit from as much planning up-front as possible to overcome their disadvantages.

I then conducted a similar sensitivity analysis for the quality of WBS variable, shown in Figure D.2.

Projects that do not use a WBS are also most likely outside the norm for the project life cycle (Pinto and Prescott, 1988; Kerzner, 2003). Note that both experience level of the team and quality of WBS are also independent variables as well as moderators. This could also add some complexity. Overall the model appears reasonable for these two variables.

Finally, I conducted a similar sensitivity analysis for the internal versus vendor variable, shown in Figure D.3. The results for internal versus vendor look fully consistent. For 100% vendor completed, the curve is flat, indicating a very small or no relationship between planning and success. This is logically consistent as for those projects the important detailed planning is done by the vendor and not by the respondent of the survey who is reporting his own planning.

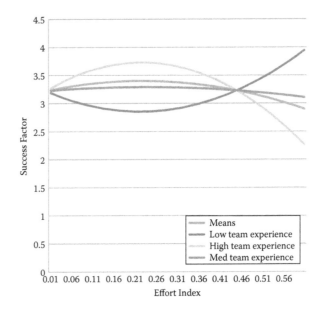

FIGURE D.1
Planning index versus success measure for the final model sensitivity analysis: experience
level of team values.

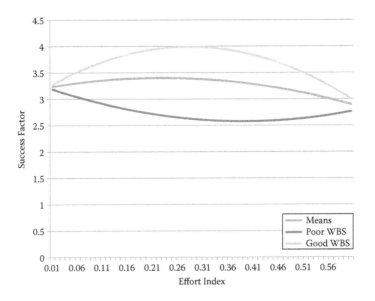

FIGURE D.2
Planning index versus success measure for the final model sensitivity analysis: border
quality of WBS values.

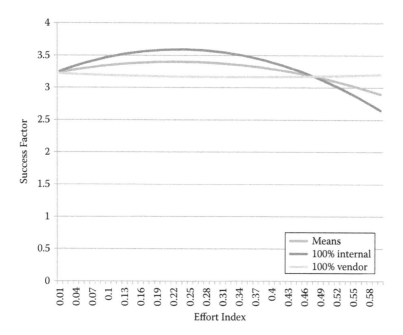

FIGURE D.3

Planning index versus success measure for the final model sensitivity analysis: border internal versus vendor values.

We can conclude the model works well within reasonable effort range from both a conceptual and an empirical standpoint. We can state that high-experience teams have the most to gain from spending optimal effort at the planning stage, as well as teams with a solid WBS. However, spending too little or too much effort can reduce the overall success impact of even the best teams or the best WBS.

References

Andersen, E.S. (1996). Warning: Activity planning is hazardous to your project's health. *International Journal of Project Management* 2(14): 89–94.

Association for Project Management (2007). *APM Body of Knowledge*. UK: The Association for Project Management. http://www.knowledge.apm.org.uk/bok-splash.

Aubry, M., Hobbs, B., and Thuillier, D. (2008). Organisational project management: An historical approach to the study of PMOs. *International Journal of Project Management* 26(1): 38–43.

Bachy, G. and Hameri, A.-P. (1997). What to be implemented at the early stage of a large-scale project. *International Journal of Project Management* 15(4): 211–218.

Bart, C. (1993). Controlling new product R&D projects. *R&D Management* 23(3): 187–198.

Besner, C. and Hobbs, B. (2006). The perceived value and potential contribution of project management practices to project success. *Project Management Journal* 37(3): 37–48.

Besner, C. and Hobbs, B. (2011). Contextualised project management practice: A cluster analysis of practices and best practices. In *10th IRNOP Research Conference, Montreal Canada*.

Beyer, W. (1987). *CRC Standard Mathematical Tables*. Boca Raton, FL: CRC Press.

Blackburn, S. (2005). *The Oxford Dictionary of Philosophy*. New York: Oxford University Press.

Blomquist, T., Hällgren, M., Nilsson, A., and Söderholm, A. (2010). Project-as-practice: In search of project management research that matters. *Project Management Journal* 41(1): 5–16.

Boehm, B. (1996). Anchoring the software process. *IEEE Software* 13(4): 73–82.

Boehm, B. (2002). Get ready for agile methods, with care. *Computer* 35(1): 64–69.

Boehm, B. and Turner, R. (2003). Observations on balancing discipline and agility. In *Proceedings of the Conference on Agile Development*, vol. 25–28, pp. 32–39.

Boynton, A.C. and Zmud, R.W. (1984). An assessment of critical success factors. *Sloan Management Review* 25(4): 17–27.

Brennan, K. (2009). *A Guide to the Business Analysis Body of Knowledge (BABOK® Guide)*. Marietta, GA: International Institute of Business Analysis.

Brown, I.T.J. (2004). Testing and extending theory in strategic information systems planning through literature analysis. *Information Resources Management Journal* 17(4): 20–48.

Brunnermeier, M.K., Papakonstantinou, F., and Parker, J.A. (2008). *An economic model of the planning fallacy* (Working Paper 14228). National Bureau of Economic Research, Cambridge.

Buehler, R., Griffin, D., and Ross, M. (1994). Exploring the planning fallacy: Why people underestimate their task completion times. *Journal of Personality and Social Psychology* 67(3): 366–381.

Carroll, S.J. and Gillen, D.J. (1984). The classical management functions: Are they really outdated? In *Academy of Management Proceedings* (00650668) (pp. 132–136).

Carroll, S.J. and Gillen, D.J. (1987). Are the classical management functions useful in describing managerial work? *Academy of Management Review* 12(1): 38–51.

Catersels, R., Helms, R.W., and Batenburg, R.S. (2010). Exploring the gap between the practical and theoretical world of ERP implementations: Results of a global survey.

In *Proceedings of IV IFIP International Conference on Research and Practical Issues of Enterprise Information Systems.*

Cerpa, N. and Verner, J.M. (2009). Why did your project fail? *Communications of the ACM* 52(12): 130–134.

Ceschi, M., Sillitti, A., Succi, G., and De Panfilis, S. (2005). Project management in plan-based and agile companies. *IEEE Software* 22(3): 21–27.

Chatzoglou, P. and Macaulay, L.A. (1996). Requirements capture and IS methodologies. *Information Systems Journal* 6(3): 209–225.

Choma, A.A. and Bhat, S. (2010). Success vs failure: What is the difference between the best and worst projects? In *Proceedings PMI Global Congress 2010,* Washington, DC.

Christenson, D. and Walker, D. (2008). Using vision as a critical success element in project management. *International Journal of Managing Projects in Business* 1(4): 611–622.

Churchill Jr, G. (1979). A paradigm for developing better measures of marketing constructs. *Journal of Marketing Research,* 64–73.

Cleland, D.I. and Gareis, R. (2006). *Global Project Management Handbook: Planning, Organizing and Controlling International Projects.* New York: McGraw-Hill.

Cleland, D.I. and Ireland, L.R. (2008). *Project Managers Handbook: Applying Best Practices Across Global Industries.* New York: McGraw-Hill.

Cohen, J. (1994). The earth is round ($p < .05$). *American Psychologist* 49(12): 997.

Collyer, S. and Warren, C.M. (2009). Project management approaches for dynamic environments. *International Journal of Project Management* 27(4): 355–364.

Collyer, S., Warren, C., Hemsley, B., and Stevens, C. (2010). Aim, fire, aim—Project planning styles in dynamic environments. *Project Management Journal* 41(4): 108–121.

Conway, J. and Lance, C. (2010). What reviewers should expect from authors regarding common method bias in organizational research. *Journal of Business and Psychology* 25(3): 325–334.

Cooke-Davies, T.J. (2002). The real success factors in projects. *International Journal of Project Management* 20(3): 185–190.

Cooper, D. and Schindler, P. (2008). *Business Research Methods.* New York: Irwin/McGraw-Hill.

Cooper, R.G., Edgett, S.J., and Kleinschmidt, E.J. (2004). Benchmarking best NPD practices - 1. *Research-Technology Management* 47(1): 31–43.

Coram, M. and Bohner, S. (2005). The impact of agile methods on software project management. In *Proceedings of the 12th IEEE International Conference and Workshops on Engineering of Computer-Based Systems* (pp. 363–370).

Covey, S. (2004). *Seven Habits of Highly Effective People.* New York: Free Press.

Crawford, L. (2000). Profiling the competent project manager. In *Project Management Research at the Turn of the Millennium: Proceedings of PMI Research Conference,* June 21–24, Paris (pp. 3–15).

Crawford, L. (2005). Senior management perceptions of project management competence. *International Journal of Project Management* 23(1): 7–16.

Crawford, L., Pollack, J., and England, D. (2006). Uncovering the trends in project management: Journal emphases over the last 10 years. *International Journal of Project Management* 24(2): 175–184.

Crawford, L.H., Hobbs, J.B., and Turner, J.R. (2004). Project categorization systems and their use in organisations: an empirical study. In *Proceedings of PMI Research Conference,* July, London.

Daly, E.B. (1977). Management of software development. *IEEE Transactions on Software Engineering* 3: 229–242.

De Vaus, D. (2002). *Analyzing Social Science Data – 50 Key Problems in Data Analysis*. London: Sage.

Deephouse, C., Mukhopadhyay, T., Goldenson, D.R., and Kellner, M.I. (1995). Software processes and project performance. *Journal of Management Information Systems* 12(3): 187–205.

Denscombe, M. (2007). *The Good Research Guide: For Small-Scale Social Research Projects*. Maidenhead, UK: Open University Press.

Dulewicz, V. and Higgs, M. (2005). Assessing leadership styles and organisational context. *Journal of Managerial Psychology* 20(2): 105–123.

Dvir, D. and Lechler, T. (2004). Plans are nothing, changing plans is everything: The impact of changes on project success. *Research Policy* 33(1): 1–15.

Dvir, D., Raz, T., and Shenhar, A. (2003). An empirical analysis of the relationship between project planning and project success. *International Journal of Project Management* 21(2): 89–95.

Dybå, T. and Dingsøyr, T. (2008). Empirical studies of agile software development: A systematic review. *Information and Software Technology* 50(9): 833–859.

Eisenhardt, K. (1989). Making fast strategic decisions in high-velocity environments. *Academy of Management Journal* 32(3): 543–576.

Ellwood, M. (2005). Time priorities for top managers. In *International Association of Time Use Researchers Annual Conference*, Halifax, Canada.

Ewusi-Mensah, K. (1997). Critical issues in abandoned information systems development projects. *Communications of the ACM* 40(9): 74–80.

Faniran, O.O., Oluwoye, J.O., and Lenard, D.J. (1998). Interactions between construction planning and influence factors. *Journal of Construction Engineering & Management* 124(4): 245.

Fellows, R. and Liu, A. (2003). *Research Methods for Construction*. Oxford and Malden, MA: Blackwell Science.

Field, A. (2009). *Discovering Statistics Using SPSS*. London: Sage.

Fischhoff, B. (1991). Value elicitation: Is there anything in there? *American Psychologist* 46(8): 835–847.

Fitzgerald, B. (1996). Formalized systems development methodologies: A critical perspective. *Information Systems Journal* 6(1): 3–23.

Flyvbjerg, B., Holm, M.S., and Buhl, S. (2002). Underestimating costs in public works projects: Error or lie? *Journal of the American Planning Association* 68(3): 279–295.

Furuyama, T., Arai, Y., and Lio, K. (1993). Fault generation model and mental stress effect analysis. In *Proceedings of the Second International Conference on Achieving Quality in Software*, October 18–20, Venice, Italy.

Gantt, H. (1910). *Work, Wages and Profit, published by The Engineering Magazine, New York, 1910; republished as Work, Wages and Profits*, Easton, PA: Hive, 1974.

Gibson, E. and Dumont, P. (1995). *Project Definition Rating Index (PDRI) for Industrial Projects; CII Research Report 113-11*. Austin, TX: The Construction Industry Institute.

Gibson, E. and Gebken, R. (2003). Design quality in pre-project planning: Applications of the project definition rating index. *Building Research and Information* 31(5): 346–356.

Gibson, E. and Pappas, M.P. (2003). *Starting Smart: Key Practices for Developing Scopes of Work for Facility Projects*. Washington, DC: National Academies Press.

Gibson, G., Wang, Y., Cho, C., and Pappas, M. (2006). What is pre-project planning, anyway? *Journal of Management in Engineering* 22(1): 35–42.

Goetz, B.E. (1949). *Management Planning and Control: A Managerial Approach to Industrial Accounting*. New York: McGraw-Hill.

Gopal, A., Mukhopadhyay, T., and Krishnan, M.S. (2002). The role of software processes and communication in offshore software development. *Communications of the ACM* 45(4): 193–200.

Gulick, L.H. (1936). *Notes on the Theory of Organization*. Gulick, L. and Urwick, L. (eds), Papers on the Science of Administration. New York Institute of Public Administration.

Hällgren, M. and Maaninen-Olsson, E. (2005). Deviations, ambiguity and uncertainty in a project-intensive organization. *Project Management Journal* 36(3): 17–26.

Hamilton, M.R. and Gibson, G.E.J. (1996). Benchmarking preproject-planning effort. *Journal of Management in Engineering* 12(2): 25–33.

Howell, D. (2007). *Statistical Methods for Psychology*. Belmont, CA: Thomson Wadsworth.

Hubbard, R. and Lindsay, R.M. (2008). Why p values are not a useful measure of evidence in statistical significance testing. *Theory & Psychology* 18(1): 69–88.

Hwang, M., Windsor, J., and Pryor, A. (2000). Building a knowledge base for MIS research: A meta-analysis of a systems success model. *Information Resources Management Journal* 13(2): 26–32.

ISO (2003). *10006: 2003. Quality management systems. Guidelines for quality management in projects*. British Standards Institution, London UK (Technical report, British Standards Institution, London UK).

Jiang, J.J., Klein, G., and Chen, H.-G. (2001). The relative influence of is project implementation policies and project leadership on eventual outcomes. *Project Management Journal* 32(3): 49–55.

Johnson, J., Boucher, K., and Connors, K. (2001). Collaborating on project success. *Software Magazine* 7(2): 1–9.

Jones, C. (1986). *Programming Productivity*. New York: McGraw-Hill.

Jorgensen, M. and Boehm, B. (2009). Software development effort estimation: Formal models or expert judgment? *IEEE Software* 26(2): 14–19.

Jorgensen, M. and Grimstad, S. (2011). The impact of irrelevant and misleading information on software development effort estimates: A randomized controlled field experiment. *IEEE Transactions on Software Engineering* 37(5): 695–707.

Jugdev, K. and Müller, R. (2005). A retrospective look at our evolving understanding of project success. *Project Management Journal* 36(4): 19–31.

Kaner, C., Falk, J., and Nguyen, H.Q. (1999). *Testing Computer Software* (2nd ed.). New York: Wiley.

Kapsali, M. (2011). *Relating in Project Networks and Innovation Systems*. Retrieved from http://ssrn.com/abstract=1969395.

Kerzner, H. (2003). *Project Management: A Systems Approach to Planning, Scheduling, and Controlling* (8th ed.). New York: Wiley.

Kerzner, H. (2009). *Project Management: A Systems Approach to Planning, Scheduling, and Controlling*, (10th ed.). New York: Wiley.

King, W.R. (1988). How effective is your information systems planning? *Long Range Planning* 21(5): 103–112.

Kirton, M. (1976). Adaptors and innovators: A description and measure. *Journal of Applied Psychology* 61(5): 622–629.

Kloppenborg, T.J., Manolis, C., and Tesch, D. (2009). Successful project sponsor behaviors during project initiation: An empirical investigation. *Journal of Managerial Issues* 21(1): 140–159.

Koontz, H. (1958). A preliminary statement of principles of planning and control. *The Journal of the Academy of Management* 1(1): 45–61.

Koontz, H. and Weihrich, H. (2006). *Essentials of Management*. New Delhi, India: McGraw-Hill Education (India) Pvt Ltd.

Koskela, J. and Abrahamsson, P. (2004). On-site customer in an XP project: Empirical results from a case study. In T. Dingsøyr (Ed.), *Software Process Improvement,* Vol. 3281 (pp. 1–11). Berlin/Heidelberg: Springer.

Koskela, L. and Howell, G. (2002). The underlying theory of project management is obsolete. *IEEE Engineering Management Review* 36(2): 22–34.

Kotter, J.P. (1999). What effective general managers really do. *Harvard Business Review* 60(6): 3–12.

Lamers, M. (2002). Do you manage a project, or what? A reply to "Do you manage work, deliverables or resources?" *International Journal of Project Management* 20(4): 325–329.

Lindvall, M., Basili, V., Boehm, B., Costa, P., Dangle, K., Shull, F., Tesoriero, R., Williams, L., and Zelkowitz, M. (2002). Empirical findings in agile methods. *Extreme Programming and Agile Methods—XP/Agile Universe 2002* 2418: 81–92.

Love, P.E.D., Edwards, D.J., and Irani, Z. (2008). Forensic project management: An exploratory examination of the causal behavior of design-induced rework. *IEEE Transactions on Engineering Management* 55(2): 234–247.

Luthans, F. (1988). Successful vs. effective real managers. *The Academy of Management Executive (1987)* 2(2): 127–132.

Magazinius, A. and Feldt, R. (2011). Confirming distortional behaviors in software cost estimation practice. In *Proceedings of the 37th EUROMICRO Conference on Software Engineering and Advanced Applications (SEAA)* (pp. 411–418).

Mankins, M. (2004). Stop wasting valuable time. *Harvard Business Review* 82(9): 58–65.

Mann, C. and Maurer, F. (2005). A case study on the impact of scrum on overtime and customer satisfaction. In *Agile Conference, 2005 Proceedings* (pp. 70–79).

Marshall, R.A. (2007). *A quantitative study of the contribution of earned value management to project success on external projects under contract.* Unpublished doctoral disseration, ESC Lille, Lille, France.

Martin, N.L., Pearson, J.M., and Furumo, K.A. (2005). IS project management: Size, complexity, practices and the project management office. In *System Sciences, 2005. HICSS '05. Proceedings of the 38th Annual Hawaii International Conference* (p. 234b).

McFarlan, F. W. (1981). Portfolio approach to information systems.. *Harvard Business Review* 59(5): 142–150.

Milosevic, D. and Patanakul, P. (2005). Standardized project management may increase development projects success. *International Journal of Project Management* 23(3): 181–192.

Mintzberg, H. (1975). The manager's job: Folklore and fact. *Harvard Business Review* 53(4): 49–61.

Mintzberg, H. (1994). *The Rise and Fall of Strategic Planning: Reconceiving Roles for Planning, Plans, Planners.* Englewood Cliffs, NJ: Prentice Hall.

Morris, P.W.G. (1998). Key issues in project management. In J.K. Pinto (Ed.), *Project Management Institute Project Management Handbook.* Newtown Square, PA: Project Management Institute.

Müller, R. and Turner, J.R. (2001). The impact of performance in project management knowledge areas on earned value results in information technology projects. *International Project Management Journal* 7(1): 44–51.

Müller, R. and Turner, J.R. (2007). Matching the project manager's leadership style to project type. *International Journal of Project Management* 25(1): 21–32.

Munns, A. and Bjeirmi, B. (1996). The role of project management in achieving project success. *International Journal of Project Management* 14(2): 81–87.

Murray, A., Bennett, N., Bentley, C, (2009). *Managing Successful Projects with Prince2.* London: Stationery Office.

Narins, P. (1999). Get better information from all your questionnaires—13 important tips to help you pretest your surveys. *SPSS Keywords Online.* Retrieved from http://www. uoguelph.ca/htm/MJResearch/ResearchProcess/PretestingTips.htm.

Nobelius, D. and Trygg, L. (2002). Stop chasing the front end process–management of the early phases in product development projects. *International Journal of Project Management* 20(5): 331–340.

Norman, G. (2010). Likert scales, levels of measurement and the "laws" of statistics. *Advances in Health Sciences Education* 15(5): 625–632.

Nunnally, J. (1978). *Psychometric Theory.* New York: McGraw-Hill.

Olomolaiye, A. (2007). *The impact of human resource management on knowledge management for performance improvements in construction organisations.* Unpublished doctoral dissertation, Glasgow Caledonian University, Glasgow, UK.

Olson, B. and Swenson, D. (2011). Overtime effects on project team effectiveness. In *The Midwest Instruction and Computing Symposium,* April, Duluth, MN.

Orpen, C. (1985). The effects of long-range planning on small business performance: A further examination. *Journal of Small Business Management* 23(1): 16–23.

Pankratz, O. and Loebbecke, C. (2011). Project managers' perception of is project success factors—A repertory grid investigation. In *ECIS 2011 Proceedings,* Vol. 170.

Pearsall, J. (1999). *The Concise Oxford Dictionary.* London: Oxford University Press.

Pinto, J.K. and Prescott, J.E. (1988). Variations in critical success factors over the stages in the project life cycle. *Journal of Management* 14(1): 5–18.

Pinto, J.K. and Prescott, J.E. (1990). Planning and tactical factors in the project implementation process. *Journal of Management Studies* 27(3): 305–327.

Pinto, J.K. and Slevin, D.P. (1988). Project success: Definitions and measurement techniques. *Project Management Journal* 19(1): 67–72.

PMI, Project Management Institute. (2013). *A Guide to the Project Management Body of Knowledge* (5th ed.). Newtown Square, PA: Project Management Institute.

Podsakoff, P., MacKenzie, S., Lee, J., and Podsakoff, N. (2003). Common method biases in behavioral research: A critical review of the literature and recommended remedies. *Journal of Applied Psychology* 88(5): 879–903.

Poon, S., Young, R., Irandoost, S., and Land, L. (2011). Re-assessing the importance of necessary or sufficient conditions of critical success factors in it project success: A fuzzy set-theoretic approach. In *ECIS 2011 Proceedings,* Vol. 176.

Posten, R.M. (1985). Preventing software requirements specification errors with IEEE 830. *IEEE Software* 2(1): 83–86.

Prabhakar, G. (2008). What is project success: A literature review. *International Journal of Business and Management* 3(8): 3–10.

Premkumar, G. and King, W.R. (1991). Assessing strategic information systems planning. *Long Range Planning* 24(5): 41–58.

Premkumar, G. and King, W.R. (1992). An empirical assessment of information systems planning and the role of information systems in organizations. *Journal of Management Information Systems* 9(2): 99–125.

Project Management Institute (PMI®) (2013). *A Guide to the Project Management Body of Knowledge* (PMBOK® Guide) (5th ed.). Newtown Square, PA: Project Management Institute.

Putnam, L.H. and Myers, W. (1997). How solved is the cost estimation problem? *IEEE Software* 14(6): 105–107.

Reel, J.S. (1999). Critical success factors in software projects. *IEEE Software* 16(3): 18–23.

Roscoe, J. (1975). *Fundamental Research Statistics for the Behavioral Sciences*. New York: Holt, Rinehart and Winston.

Rosenberg, D. and Scott, K. (1999). *Use Case Driven Object Modeling with UML: A Practical Approach*. Reading, MA: Addison-Wesley.

Royall, R.M. (1986). The effect of sample size on the meaning of significance tests. *The American Statistician* 40(4): 313–315.

Salomo, S., Weise, J., and Gemünden, H. (2007). NPD planning activities and innovation performance: The mediating role of process management and the moderating effect of product innovativeness. *Journal of Product Innovation Management* 24(4): 285–302.

Schultz, R.L., Slevin, D.P., and Pinto, J.K. (1987). Strategy and tactics in a process model of project implementation. *Interfaces* 17(3): 34–46.

Scott-Young, C. and Samson, D. (2008). Project success and project team management: Evidence from capital projects in the process industries. *Journal of Operations Management* 26(6): 749–766.

Segars, A.H. and Grover, V. (1998). Strategic information systems planning success: An investigation of the construct and its measurement. *MIS Quarterly* 22(2): 139–163.

Serrador, P. (2013). The impact of planning on project success: A literature review. *Journal of Modern Project Management* 1(2): 28–39.

Serrador, P. and Pinto, J.K. (2014, Working Paper). Does Agile Work? A Quantitative Analysis of Project Success.

Serrador, P. and Turner, J R. (2013). The impact of the planning phase on project success. In *Proceedings of IRNOP 2013*, Oslo, Norway.

Serrador, P. and Turner, J.R. (2014). The relationship between project success and project efficiency. *Procedia - Social and Behavioral Sciences* 119: 75–84.

Sessions, R. (2009). *The IT complexity crisis: Danger and opportunity*. ObjectWatch, Inc. Retrieved from http://www.objectwatch.com/whitepapers/ITComplexity WhitePaper.pdf.

Sharma, S., Durand, R., and Gur-Arie, O. (1981). Identification and analysis of moderator variables. *Journal of Marketing Research* 18: 291–300.

Shehu, Z. and Akintoye, A. (2009). The critical success factors for effective programme management: A pragmatic approach. *The Built & Human Environment Review* 2: 1–24.

Shenhar, A.J. (2001). One size does not fit all projects: Exploring classical contingency domains. *Management Science* 47(3): 394–414.

Shenhar, A.J., Dvir, D., Levy, O., and Maltz, A. C. (2001). Project success: A multidimensional strategic concept. *Long Range Planning* 34(6): 699–725.

Shenhar, A.J., Levy, O., and Dvir, B. (1997). Mapping the dimensions of project success. *Project Management Journal* 28(2): 5–9.

Shenhar, A.J., Tishler, A., Dvir, D., Lipovetsky, S., and Lechler, T. (2002). Refining the search for project success factors: A multivariate typological approach. *R&D Management* 32(2): 111–126.

Smits, H. (2006). *5 levels of agile planning: From enterprise product vision to team stand-up*. Rally Software Development Corporation. Retrieved from http://www.rallydev.com/downloads/document/2-five-levels-of-agile-planning-from-enterprise-product-vision-to-team-stand-up.html.

Standish Group, The (2011). *CHAOS Manifesto 2011.* The Standish Group. Retrieved from http://standishgroup.com/newsroom/chaos_manifesto_2011.php

Tabachnick, B. and Fidell, L. (1989). *Using Multivariate Statistics.* New York: Harper and Row.

Tausworthe, R.C. (1980). The work breakdown structure in software project management. *Journal of Systems and Software* 1:181–186.

Teague, J.A. and Cooke-Davies, T.J. (2007). Developing organizational capability: Pointers and pitfalls. In *Proceedings PMI Global Congress 2007–EMEA,* Budapest, Hungary.

Teddlie, C. and Tashakkori, A. (2009). *Foundations of Mixed Methods Research: Integrating Quantitative and Qualitative Approaches in the Social and Behavioral Sciences.* Thousand Oaks, CA: Sage.

Thomas, M., Jacques, P.H., Adams, J.R., and Kihneman-Woote, J. (2008). Developing an effective project: Planning and team building combined. *Project Management Journal* 39(4): 105–113.

Tichy, L. and Bascom, T. (2008). The business end of IT project failure. *Mortgage Banking* 68(6): 28–35.

Trochim, W. (2006). Qualitative measures. *Research Methods Knowledge Base.* http://www.socialresearchmethods.net/kb/qual.php

Tullett, A.D. (1996). The thinking style of the managers of multiple projects: Implications for problem solving when managing change. *International Journal of Project Management* 14(5): 281–287.

Turner, J.R. (1999). *The Handbook of Project-Based Management: Improving the Processes for Achieving Strategic Objectives.* London: McGraw-Hill.

Turner, J.R. and Cochrane, R.A. (1993). Goals-and-methods matrix: Coping with projects with ill defined goals and/or methods of achieving them. *International Journal of Project Management* 11(2): 93–102.

Turner, J.R., Huemann, M., Anbari, F.T., and Bredillet, C.N. (2010). *Perspectives on Projects.* London and New York: Routledge.

Turner, J.R. and Müller, R. (2003). On the nature of the project as a temporary organization. *International Journal of Project Management* 21(1): 1–8.

Turner, J.R. and Müller, R. (2005). The project manager's leadership style as a success factor on projects: A literature review. *Project Management Journal* 36(2): 49–61.

Umble, E.J., Haft, R.R., and Umble, M. (2003). Enterprise resource planning: Implementation procedures and critical success factors. *European Journal of Operational Research* 146(2): 241–257.

Van Genuchten, M. (1991). Why is software late? An empirical study of reasons for delay in software development. *IEEE Transactions on Software Engineering* 17(6): 582–590.

van Marrewijk, A., Clegg, S.R., Pitsis, T.S., and Veenswijk, M. (2008). Managing public-private megaprojects: Paradoxes, complexity, and project design. *International Journal of Project Management* 26(6): 591–600.

Wang, Y.-R. and Gibson, G.E. (2008). A study of preproject planning and project success using ANN and regression models. In *The 25th International Symposium on Automation and Robotics in Construction, ISARC-2008* (pp. 688–696).

White, D. and Fortune, J. (2002). Current practice in project management: An empirical study. *International Journal of Project Management* 20(1): 1–11.

Wideman, M. (2000). *Managing the development of building projects for better results.* Retrieved from www.maxwideman.com.

Wonnacott, T., and Wonnacott, R. (1990). *Introductory Statistics for Business and Economics* (4th ed.). New York: John Wiley & Sons.

Yeo, K.T. (2002). Critical failure factors in information system projects. *International Journal of Project Management* 20(3): 241–246.

Yetton, P., Martin, A., Sharma, R., and Johnston, K. (2000). A model of information systems development project performance. *Information Systems Journal* 10(4): 263–289.

Zwikael, O. (2009). The relative importance of the PMBOK® Guide's nine Knowledge Areas during project planning. *Project Management Journal* 40(4): 94–103.

Zwikael, O. and Globerson, S. (2006). Benchmarking of project planning and success in selected industries. *Benchmarking: An International Journal* 13(6): 688–700.

Index

Milton Keynes UK
Ingram Content Group UK Ltd.
UKHW031131141024
449569UK00006B/270